THE
DELUSIONS OF
EVOLUTION

Why mere accidents of chemistry cannot account
for Life, Mind or the modern evidence for Life
after death and why the ideology of
materialism distorts science

Daurie Laurence

With grateful acknowledgement and dedication to all those, whether materialist or idealist in outlook, who have sought to expand human knowledge.

Copyright © 2013 Daurie Laurence

All rights reserved. No part of this publication may be reproduced or transmitted in any form or by any means, electronic or mechanical including photocopying, recording or any information storage or retrieval system, without prior permission in writing from the publisher.

The right of Daurie Laurence to be identified as the author of this work has been asserted by him in accordance with the Copyright, Designs and Patents Act 1988

First published in Great Britain in 2013 by
The Choir Press

ISBN 978-1-909300-11-8

Contents

Introduction	*page* v
Prologue: Magical Living World	ix
1. Two Kinds of Evolution: Random and Intelligent	1
2. Idealism: Life is Intelligent – The Classical View	21
3. Materialism: Life is an Accident – But is This Really True?	35
4. Codes: Cannot Arise by Accident – It's Impossible	48
5. Materialist–Science: Tries to Abolish the Soul and Life After Death	69
6. Living Cells: Like Chemical Factories, Only More Complicated	77
7. Evolution: Intelligent Unfoldment or Random Fluke?	86
8. People: Accidental Bio-Robot Computers or Living Souls?	103
9. Court Says: Intelligent Design 'May Be True'	117
10. Head to Head: Unintelligent Design versus Intelligent Design	131
11. Material–Science: Brilliantly Explains How Things Work	150
12. Aristotle's Four Causes: Without them Nothing Makes Sense	170
13. Spiritual–Science: The Soul persists, there is Life After Death	186
14. Materialism: A Failed Hypothesis and a Science Blocker?	205
15. Magical Living World: Mindless Mystery or Blazing Mystery?	227

Introduction

To many of us, our physical lives, our amazing immaterial minds and the Living World as a whole, are astonishing things, miraculous even. Yet modern, materialist theories of life, mind and evolution argue that the amazing unfoldment of the Living World, across deep time, that the fossils make visible, is simply a very long running 'accident'.

At the same time, we now know, as Darwin did not, that Living Cells, here from the very beginning, 3.8 billion years ago, are, incredible as it sounds, like micro-miniaturized chemical factories and that the DNA Coding software that helps to run them is 'like a computer program but far, far more advanced than any software we have ever created'.[1] We also know that chemical factories, large or micro-miniaturized, never arise 'by accident', and that computer programs – whether simple or advanced – never write themselves.

Is it, therefore, really true that such incredible biological complexity, here from the beginning, arose and continues to arise, *this very second*, every second, 24/7, 7/52, by unintelligent and purposeless processes? Is there any real evidence that such claims are actually true? Or are they, primarily, the assertions of a science that has been captured by a *materialist* ideology which holds that the entire cosmos is nothing more than a vast 'accident' – and, therefore, that the Living World 'has to be' too, regardless of all the evidence for *intelligence and purpose* in Living Nature that makes such a claim unreasonable.

Increasingly, though, even non-religious intellectuals are returning to the classical, pre-Darwinian view that Life and Mind cannot be fully accounted for in terms of material laws and chemicals alone. Life-long atheist Antony Flew decided that the digital, DNA CODED information in the Cell, among other things, could not be explained materialistically, and wrote *There is A God: How the world's most notorious atheist changed his mind* (2009). Richard Milton, a science journalist, a religious agnostic, wrote *Shattering the Myths of Darwinism* (1992) pointing out that the Darwinian theory of evolution has become more 'an act of faith rather than a functioning science.'

The philosopher Thomas Nagel, who has no religious position, has written *Mind and Cosmos: Why The Materialist Neo-Darwinian Conception of Nature is Almost Certainly False* (2012). He argues that mind can neither be reduced to brain chemicals, nor can it be fully

---

[1] Gates, B. 1995. The Road Ahead. Viking.

explained materialistically. While noting that 'physico-chemical reductionism in biology is the orthodox view, and any resistance to it is regarded as not only scientifically but politically incorrect,' he says that for a long time he has 'found the materialist account of how we and our fellow organisms came to exist hard to believe, including the standard version of how the evolutionary process works.' He says, 'The more details we learn about the chemical basis of life and the intricacy of the genetic code, the more unbelievable the standard historical account becomes.' He mentions R. Dawkins book *The Blind Watchmaker* as an example of the standard account.

It seems to him that the current scientific orthodoxy 'about the cosmic order is the product of governing [materialist] assumptions that are unsupported [by the evidence], and that it flies in the face of common sense.' He says, 'The world is an astonishing place, and the idea that we have ... the basic tools needed to understand it is no more credible now than it was in Aristotle's day.' Nagel questions the prevailing materialistic scientific paradigm, and, by contrast with Dawkins, upholds the classical view that Life and Mind almost certainly cannot be fully explained in terms of purely physical matter and its laws alone.

There is a dispute in science and philosophy, today, and in the wider culture, between those who consider that Mind and Life can be fully explained (or will, one day, be fully explained), in terms of nothing more than the known laws of physics and chemistry (the materialist view), and those who argue that there is more to life and mind, and their unfoldment across vast time, than physics and chemistry. There are other more subtle laws and processes at work that involve intelligence, meaning and purpose – Soul Forces, Life Forces, Spiritual Forces and so on. The second view, which I attempt to present in this book, is the non-materialist or the idealist view. Obviously, anyone with any kind of religious or spiritual belief is on the non-materialist or idealist side of the argument.

The picture is confused, however, because many scientists, while believing in the Divine or some kind of Spiritual reality behind Life and Mind and outer Nature, do, at the same time, believe in the Darwinian theory of evolution by means of the natural selection – the survival in the wild – of allegedly *random* genetic variations.

This is a problematic, even self-contradictory, stance because the whole purpose of Darwin's theory of evolution was to provide an entirely materialist or non-super-naturalist (ie god-or-intelligence-not-needed) explanation of Life and Mind on earth. It was not, in its essence, an explanation intended to appeal to religious and spiritual believers! The whole point of materialism (sometimes also referred to as naturalism) and materialist explanations of life and mind is that they deny all things religious and spiritual. They hold that only physical matter is real and, therefore, that Nature in its totality is mindless and mono-dimensional as opposed to intelligently multi-dimensional and including within in itself many subtle, unseen, higher order laws, forces and agencies that are not detectable by the purely physical senses or their mechanical extensions but whose existence can be *rationally inferred* or directly cognized by intuitive, extrasensory or spiritual means.

At the heart of this controversial topic, in which science and religion are often presented as being in conflict with each other, are key questions as to the true nature of Nature, the true nature of Evolution and what does the word evolution, when used in relation to the Living World, actually mean? What we think Evolution is depends very much on what we consider to be the true nature of Nature itself.

Is 'Nature,' in the sense of 'All That Exists,' just chemicals and nothing more, as materialist thinkers like R. Dawkins holds and the late S.J. Gould and F. Crick maintained, or does it contain subtle, intelligent and purposeful 'super' or 'higher' Natural components, as most scientists prior to Darwin, and many since, have held. If total Nature really is just matter alone, just 'ordinary' (non 'super') Natural physics and chemistry, and that is all that exists then science and religion are bound to conflict because if only physics and chemistry are real then, by definition, all talk of subtle, higher formative and evolutionary laws, forces and processes, all talk of 'super' Nature, of the religious, the Spiritual and the Soulful becomes meaningless.

Few people, apart from those who take the christian bible very literally, (sometimes referred to as 'young earth creationists'), dispute that Life on earth is of immense antiquity and that it has vastly changed, over vast time. The dispute, however, is as to the nature of those changes or evolution. How did they happen and what drove them? Were, and are, any intelligent and purposeful laws, forces and processes or agencies involved or were they driven purely by the known laws of Life Forceless and Mindless physics and chemistry in combination, it is alleged, with millions of random hypothetical events or 'accidents' as materialist thinkers like Dawkins, Crick and Gould have argued?

Materialist thinking science argues that chance chemical events (for example, random genetic mutations) explain all. But, even though, from the media and general schooling, not to mention the widely disseminated writings of R. Dawkins, one wouldn't know it, this is much more a *materialist* hypothesis about Life and Mind rather than a substantiated fact (as pointed out, for example, by M. Behe in *The Edge of Evolution: The Search for the Limits of Darwinism*) and it doesn't, in any case, explain, (i) how Living Cells of stunning complexity arose in the first place, (ii) how genes arose in the first place, (iii) how astonishingly complex, digital DNA Coding could arise 'by accident,' and in defiance of the laws of mathematics and information theory (as explained by S. Meyer in *Signature in the Cell: DNA and the Evidence for Intelligent Design*), (iv) what 'Life' actually is, (v) what Mind, Consciousness and Sentiency actually are (as Thomas Nagel points out in Mind and Cosmos: *Why the Materialist neo-Darwinian Conception of Nature is Almost Certainly False*).

In summary, the four basic positions on evolution or Life's unfoldment across vast time are:

1. Those who use the christian bible to argue that the world is just 6000 years old and was literally made in 6 days. They are often referred to by the mainstream media and scientific community as creationists or young-earth creationists.
2. Other christians hold that the bible is metaphorical, that 'God or Supreme Being is the Creator', and that His/Her/Its way of creating (and/or evolving) was (a) via the Big Bang and (b) via Darwin and Wallace's proposed mechanisms of natural selection of random (or deliberately induced) genetic variations. These *theistic* thinking Darwinians are not materialists (they believe in the 'super' Natural and in the multi-dimensionality of the totality), but, when it comes to the science of natural history, they are not always easy to distinguish from materialist thinkers like Dawkins, Crick and Gould.

There are many scientists in this second category, such as, famously, Francis Collins, head of the Human Genome Project and author of *The Language of God: A Scientist Presents Evidence for Belief*. They believe in God or 'super' Nature but, at the same time, they believe that Darwin discovered God's way of creating – by apparently random processes across vast time. These thinkers, unlike Richard Dawkins, are theists but, at the same time, in harmony with Darwin and Dawkins, they do not believe in 'intelligent design' or that scientists are capable of detecting the signatures of intelligent versus random causation in Living Nature.

Really, though, their position is not only a subversion of Darwinism (which is, as Richard Dawkins points out, an attempt to provide an entirely materialistic or intelligence-not-needed explanation of Life and Mind on earth), but it's also self-contradictory. This is because either the divine or intelligent laws and forces of some kind or kinds are involved in the arising of the Living World, and, if so, scientists are capable of detecting them (as modern theorists of intelligent design such as Michael Behe, William Dembski and Stephen Meyer, among others, argue), or there are no such intelligent laws and processes and, therefore, there are no signs of intelligence or 'intelligent design' there to be detected - which is the materialist view point of thinkers like Darwin, Crick and Dawkins, who argue that Living things may 'appear' to be 'designed' but they are not. The all-pervasive appearances of intelligence, purpose and design in Living Nature are illusions and it is just wishful thinking to believe otherwise – the random chances of physics and chemistry are sufficient to explain all including very consciousness, sentiency, subjectivity and Mind; although, it must be said, this materialist view begs the question as to where the amazing laws of physics and chemistry and the clever properties of chemical matter came from in the first place.

3. Then there are those christians, and others, who are not 'young earth' or 'six day creationists' but who, at the same time, are not Darwinians because they believe (1) that 'accidents of chemistry' cannot explain Life or intelligent Mind and (2) that modern science and scientists (not the bible or any other religious books) are capable of (a) detecting and (b) have detected evidence of intelligence and intelligent design in many aspects of nature and the Living World – such as the *specified* complexity of the DNA Coding (Stephen Meyer's *Signature in the Cell*) and *irreducible* complexity (biochemist, Michael Behe's *Darwin's Black Box*).

4. Finally, there are the materialist thinkers, like Richard Dawkins (*The Selfish Gene, The Blind Watchmaker* etc), Carl Sagan (*The Demon Haunted World*), Francis Crick and others, who believe that 'only matter is real' and that by an improbable series of accidents of chemistry (a) Life and Mind arose in the first place and (b) they randomly, without any inherent purpose or intelligence, 'evolved' to become today's living world.

Taking a keen interest in these controversial matters, two friends, Oliver and Alisha explore the idea, while on their walk in some beautiful countryside, that evolution, of some kind, over millions of years, is clearly a fact. But, is it a purposeful and intelligent fact, or a random one?

In this book they focus principally on Life's intelligent origins and in a follow up book, *Evolution, Truth and Delusion: Why Life is More Than Chemistry*, they focus somewhat more on its unfoldment through deep time. They are inclined to conclude that the Living World, and we, as part of it, are not the incredibly clever and 24/7 dynamic results of 'accidental' chemistry but are intelligently evolved unfoldments. Life is a magical mystery, a miracle, not an accident. Today, it is science, as much as religion and spirituality, that confirms our nearly universal human intuitions that there is more to Life than meets the overly materialistic eye.

I wanted to say something on this complex and controversial topic because I have thought for a long time that randomly tumbling components of anything, be they watch parts, computer parts or life's chemicals, would never build anything useful. To claim otherwise seems anti-empirical and illogical. It flies in the face of all that we empirically know of the different kinds of causation – intelligent and random – and the different kinds of effects that they produce. Yet, the whole materialist case is built on the assumption and firm belief that random chemical events are highly creative and evolutionary and that existence is, fundamentally, an 'accident.'

For the materialist perspective so dominant in science today to seem correct it must (a) claim that Life and Mind are 'accidents of chemistry' (a chemistry with amazing properties and laws that, materialism argues, is also a random fluke and nothing meaningful), (b) it must ignore the calculations of those, such as Sir Frederick Hoyle, William Dembski and various others, who have pointed out that the laws of mathematics forbid the arising of Life, or Life's Codes, 'by chance', no matter how much time is allowed, and (c) it must quietly ignore or actively discount large bodies of human testimony – spiritual, religious, out of body experiences, near death experiences, telepathy, medium channelled communications, spiritual healing etc – that there is more to Life and Mind than random chance, that there are spiritual elements to reality, that human Consciousness and individuality do continue after death and so on.

I was also motivated to write by frustration with the materialist and New Atheist thinking that seems to have such a grip on our scientific and mainstream culture. The world-view that claims that 'science has shown' Life and the Living World to be 'nothing special,' just meaningless evolutionary 'accidents'. This is a stance that regards science and the philosophy of materialism as one and the same thing. True science, though, is not a dogmatic ideology or a particular world view but rather many and varied methods for investigating the nature of reality systematically. Some scientists are philosophical materialists, matter before mind thinkers, others are philosophical idealists, mind before matter thinkers, but the ultimate nature of reality is not a matter of philosophy or belief but of what is actually true.

People can be very passionate about their beliefs so I hope that my efforts to point out the failings of the materialist world view are not too strident. I am passionate about this subject, because what is more important that the true nature of reality? Is it really all a big 'accident' with the only meanings being the ones that we make up as thinkers like Dawkins argue? Or are Nature and Existence intrinsically intelligent and meaningful? Our collective decisions about this have powerful implications because they affect so many things, not least our relations with each other and with the Living World as a whole.

Included, in this discussion about the nature of Life and its origins, is a section on some of the modern evidence for Life after death – based on more than a century's research – and the evidence that human consciousness is not, ultimately, tied to the body. This modern knowledge contradicts the materialist idea that Consciousness and Mind are produced by the chemical body–brain rather than merely transmitted by it into this physical dimension.

It is, in fact, the unjustified insistence that we have already successfully explained Life on Earth as an accident of Life–Forceless, 'dead', matter, which gives false support to the materialist idea that Life, Mind and Consciousness cannot exist beyond the 'death' of the chemical body. This is because, in materialist thinking, an 'accidental' brain gives rise to an 'accidental' Mind – although no one has a clue how.

The materialist view has many social and emotional implications but here's just one: if, as materialism claims, consciousness does not survive the death of the body, because Mind is supposedly some kind of weird, carbon based, mineral accident then suicide becomes an attractive option for those who are sad and troubled:

> "Why not end it all? After all, materialist thought teaches, it really will be the end."

But, if death is not the end, then the excarnated person, having taken this radical and irrevocable step, may feel great remorse for bringing tremendous sadness to those left behind and for the waste of a physical Life time. It is also possible that many people have felt despair, turned to drugs, even resorted to the extreme of suicide partly *because* of the pessimism of the materialist world view in which Life and All that Exists become meaningless and accidental.

Of course, if the materialist world view is true then death really is the end. Some might wish this but holistic, multi–dimensional reality just does not seem to be constructed in this way. As in the film *Groundhog*

Day it looks as though we cannot destroy ourselves as Minds and Souls but only as physical Bodies. Either way, the issue of the truth or otherwise of the survival of Consciousness and individual Mind beyond physical death is a really important one.

I also think it is relevant to our understanding of evolution because if death is not the end of the individual mind and Soul then the materialist, 'accidental particles to accidental people' form of evolution that holds that all we are is random atoms of Carbon, Oxygen, Hydrogen etc that randomly became associated, billions of years ago, into Living Cells of stunning complexity and eventually into the extremely intelligent minds and forms of a Plato, Newton or Einstein starts to seem not only unreasonable but unnecessary.

Prologue

Magical Living World

The fast flowing river sparkles in the early morning sunshine. A heron is fishing, statuesque, amazingly still. Enjoying the sounds, the pleasant pull of the river, I watch the magnificent bird earning a Living, plunging beak, scrap of silvery flapping fish, a fisher-bird's feast. In the grassy meadow, next to the river, red-brown cattle graze upon the sweet, grassy lushness. Glossy black crows hop-flap and peck at this and that on the green ground, their raven energy and intelligence palpable. There is more to them than meets the eye, these magical, strange birds, mediating between Earth and Sky, outer and inner, the obvious and the subtle, the seen and the unseen. They take off, moved by some fresh mischief. On the other side of the valley, deer move swiftly, in graceful leaps and bounds from meadow to wood. The river flows and sparkles – magical water, fabulous H_2O, sacred, liquid mineral, of amazing properties that make all Life possible, quenching and purifying. Water of Life.

A gentle breeze whispers to the grass, to the trees with softly waving leaves, to all the beings of the valley. Rustic sounds, near and far. Wood pigeons, raucous crows, murmuring cattle; the call of a distant rooster. The Living World of the plants, animals and people ... is amazing.

1 Two Kinds of Evolution: Random and Intelligent

Everything evolves – up, down or sideways. River valleys and landscapes evolve randomly and unintelligently. Cars, TVs and computers evolve intelligently. Which kind of evolution best describes the unfoldment of Living Nature across vast time?

My friend caught up with me and we walked on. After a few minutes continuing on by the sparkling river, I said: 'Ali, do you believe in Darwin's theory of evolution?'

'That's an interesting question. Why?'

'Because, after a lot of thought, I've come to the conclusion that it just doesn't make sense to think of the Living World as nothing more than a very long running 'chemical accident', literally a billions year long running *accidental chemical reaction*'.

'Surely no one thinks that. That sounds absurd.'

'I agree, it does, but for many people today, including me at one time, the famous Miller–Urey experiment in 1953, in combination with Darwin's theory of evolution by the natural selection of random chemical variations, is taken to be evidence that we and the entire Living World are literally, ongoing, 'living accidents'. The idea being that very Life itself is just some weird form of 'accidental chemistry' – with no Life or Soul Forces or any kind of motivating purpose or driving intelligence anywhere involved. This is the view, for example, made famous in books like the *Selfish Gene* and *The Blind Watchmaker* by R. Dawkins.'

'Are you saying that Dawkins thinks he, you, we and everyone are accidental chemical reactions?' said Ali incredulously.

'Look, I don't know whether he literally thinks that but that is the implication of the Darwinian–materialist world view.'

'Why though?'

'Because it assumes everything that exists, the Cosmos as a whole, to be an 'accident', that there is no purpose, intelligence or spiritual Nature to it, so what else can Life on Earth be but some weird form of accidental chemistry.'

Alisha looked troubled by my words. She said:

'That's a pretty bleak view of what Life is, isn't it?'

'I agree, but it is the materialist world view. Existence is considered to be meaningless and accidental and, obviously, if that were actually

1

true, it would mean that Life on Earth is too. For many, Darwin's theory of evolution is taken to be good evidence for that belief, a way to explain Life's unfoldment across deep time as a blind, unintelligent and wholly accidental process. The well known zoologist, Richard Dawkins, certainly thinks so. He has even said that anyone who doubts the Darwinian version of evolution is either "ignorant, stupid or insane".'

'That's a bit steep.'

'Look, I think he's quite right to believe in evolution, of *some kind*, but disagree with his view that it's a totally *random* process. I think that's a mistake and, if you're willing, I'd like to explain to you why I believe so.'

'OK, give it a go.'

'Thanks, Ali. I'll do my best to give you my honest take on it.'

We continued our walk, and, with the river next to us and the animals in the neighboring fields and meadows, I began to explain my ideas.

'The issue for me, Alisha, is not whether or not evolution takes place but the nature of that evolution. You see, evolution is everywhere. Our cars, TVs, and computers all evolve, *intelligently*. This landscape has also evolved, *randomly*. And, the Living World has evolved massively since the first microbial Life Forms appeared around 3.8 billion years ago.'

'You seem to be implying that there are two different kinds of evolution?'

'Yes, and ironically they describe two very different and more or less opposed processes – *random* evolution describes *unintelligent* unfoldments through time; for example, the evolution of landscapes, like this beautiful river valley, shaped by the sparkling river. All it requires is 'free' energy from the Sun, to drive the weather + gravity + the random weathering actions of wind, rain, frost, snow, etc, all accomplished by relatively simple accidents of chemistry. Just chemicals (earth, rocks, water) plus chance effects are fully sufficient to explain the random evolution of this river valley.

'On the other hand, we see progressive, *intelligent* evolution in technology, for example. Think of the amazing evolution of cars in just a few decades, Ford Model T to modern Ford Galaxy, or the Wright brothers' first flight to the first moon landing. By contrast with random, entropic, 'downhill' evolution, intelligent, *anti entropic*, 'uphill' evolution of the kind we see in technology requires not only Energy + Chemicals, but also the input of a lot of intelligence and *intelligent information,* that is ideas and instructions, as, for example, seen in Life's digital software, the DNA CODING. Uphill, intelligent evolution, unlike the random, downhill, evolution of a river valley or other landscape, does not benefit from 'accidents' – random, grinding, shattering and dissolving processes, erosion, corrosion, forest fires and such like.

Chemicals (metals, plastics etc) + Intelligent Information = *intelligently* evolved TVs, Cars and Computers.

Chemicals (metals, plastics etc) + Random Accidents = Car crashes and spilt coffee on the computer keyboard, not Cars or Computers.

'My argument, Ali, is that, by contrast with the Darwinian or the materialist hypothesis, the 'accidents-build-complex-Life-Forms' theory of random evolution:

"Life-Forceless Chemicals alone + Chemical Accidents alone did *not* lead to the Living World, nor do they come close to describing what it is and its amazing, 24/7 functioning accurately."

'I think it is much more reasonable to adopt the classical, pre-Darwinian view and to hypothesize that:

"Chemicals + *a lot* of Intelligent Information + Intelligent Life and Soul Forces led to the Living World, and sustain it, right now, continuously, 24/7, 7/52, for millions of years past and, for millions more years to come."

'Living organisms operate, 24/7 by way of *information*, provided in part by the DNA CODE which consists in nothing but intelligent, meaningful instructions. The CODE itself, like all codes, is a mental artifact. It may be written in deoxyribonucleic acid 'ink' but it is no more 'made of' deoxyribonucleic acid than COBOL or BASIC or Morse CODE are 'made of' whatever inks or media they are written in.

'The deoxyribonucleic acid is merely the carrier for the intelligent instructions, purposes and functions that are embodied in the code itself. Modern probability and information theory make it clear that computer type codes and software, of which the DNA CODING in our genomes is the *most sophisticated example known to humanity,* can never arise 'by accident,' by mindless, unintelligent processes. This is not merely highly improbable but just about possible, it's impossible. As Werner Gitt, professor of information theory says:

'The basic flaw of all [materialistic] evolutionary views is the origin of the information in living beings. *It has never been shown that a CODING system and semantic [meaningful, purposeful] information could originate by itself... The information theorems predict that this will never be possible.* A purely material origin of life is thus [ruled out]'[2]

'The materialist view of evolution as an entirely blind and unintelligent process, scientists are starting to realize, is impossible. This should not come as a surprise because the idea that Life is a long running chemical 'accident' was always an extraordinary claim given that there *has never been any evidence* that 'accidents' build or evolve anything clever. It is simply a materialist assumption, a Darwinian dogma that 'accidental chemistry' is supremely creative. An idea which derives from the materialist hypothesis that the entire Cosmos is an 'accident' and, therefore, Life 'has to be' too.

'Yet the Life Forms are the most complex mechanisms or 'machines' known to humanity. The animals in these fields and our own bodies make *anything* humanity makes, even our space stations or supercomputers, look like wooden spoons in comparison.'

'And your point is that supercomputers and space stations never arise without the input of *intelligently* informing forces, let alone by evolutionary 'accident'?'

'Yes, exactly. The idea that the amazing, living, breathing, breeding Soul–Life–Form Beings of the plants, animals and people are the incredibly clever and dynamic results of *accidental chemistry* is an unreasonable claim and there is no evidence for it. The real reason that so many people believe in materialistic theories of the origins of Life, and its subsequent evolution through vast time, is a form of *negative,* not positive evidence.'

'Like?'

'Like the fact that the *intelligently* causal, multi-dimensional laws, forces and processes behind and giving rise, second by second, to outer, Living Nature cannot be physically seen but can only be rationally inferred or intuited, which modern materialistic science and biology have not been willing to do.'

'Isn't that an oversimplification?'

'Look, I've put it simplistically but Darwinism was born partly in a reaction to outdated Biblical literalism in the Christian West and the

---

[2] Gitt, W. 2006. *In the Beginning Was Information.* Master Books, p. 124. Emphasis and text in square brackets added.

misguided attempts, in the previous centuries – especially at the dawn of the scientific age, starting around the 1500s – to treat the primary western religious text, the bible, not only as a source of religious and spiritual teachings, but also as a literal source of physical–scientific truths – such as trying to use it to date the age of the earth to just 6000 years ago. The creation story that goes: 'God or Supreme Intelligent Beingness said, *"Let Existence Be!"*, and so it was,' while it may be spiritually and poetically true, gives us no clue as to how the process was actually done or continues to be done, right now, day after day, week after week, for millions of years.'

'So Darwin and his Victorian contemporaries wanted to come up with an explanation of the Living World that made sense in physical–science terms?'

'Exactly.'

'However, you don't entirely agree with the explanation that he came up with?'

'That's right.'

'But you're not doubting that some form of evolution takes place, are you, Oliver? I mean – the fossils are there, aren't they?'

'Yes, they are. It is clear that evolution of some kind is real. I'm just saying that to explain Life's unfoldment across vast time, whatever its precise drivers and mechanisms, as a purely random or no-intelligence-involved process doesn't work – despite what we've all been told, despite R. Dawkins' many books on the subject, despite what many of us have believed for so long now – for generations in fact. The crucial point is that there's *nothing* in the Laws of *Life-Forceless*, 'dead', chemistry *alone* to cause random atoms of Oxygen, Carbon, Hydrogen and Nitrogen to *'accidentally'* become incredible *Living, Breathing, Breeding Organisms* any more than there's anything in 'un-organ-ized' copper ore, iron ore and crude oils in the ground to cause them to 'accidentally' mine and refine themselves out of the ground and accidentally become cars, TVs, computers etc – devices which are all vastly simpler than any single Living Cell here from the beginning of Life on Earth 3.8 billion years ago. Yet such a claim is at the heart of all purely *materialistic* theories of evolution, of which Darwinism is the most famous.'

'What do you mean by purely materialistic theories?'

'I mean ones where there are no *intelligent* Life Forces or Soul Forces involved, no *intelligently* causal laws and processes in a multi-dimensional, Holistic Nature and everything, right up to and including the entire universe, is considered to be meaningless, random and chance based, never intelligent or purposeful. According to such materialist hypotheses chemical matter is also an 'accident' and considered to be a *meaningless,* 'it's-not-clever-at-all' phenomenon.

'My difficulty with the materialist view, Ali, is how could amazingly *intelligent* mechanisms, like the Life Forms, and, as importantly, our very *Minds* and the immaterial Soul quality or spiritual *essence* that we call *intelligence* arise 'by accident' and from such, allegedly, *mindless and unintelligent* sources? It doesn't make logical or scientific sense, especially when all our experience teaches us that clever mechanisms *only ever* have *intelligent* causes, *never* chance based ones. Our TVs and computers are *much simpler* than the Life Forms, but they don't arise 'by chance'. They don't evolve by us randomly throwing their parts around.'

'Are you a creationist then?'

'It depends what you mean by a creationist. If 'creationism' means an anthropomorphic god with a beard surrounded by angels-on-harps then definitely not. But if creationism means that I don't believe Existence or our Living World to be the incredibly dynamic and clever result of utterly unintelligent processes, of 'accidental chemistry', then I'm a creationist.

'I think that Existence, of which the physically visible Universe is but one dimension, the outer and most objective dimension, is an *intelligent* phenomenon, that the *whole thing* is literally *woven* out of *intelligence*, from quarks to galaxies all the way up and all the way down. It's *all* an expression of consciousness, meaning and purpose or *sat–chit–ananda* as the Yogis say: 'truth–consciousness–bliss'.'

'So you're a kind of Nature Mystic or Panentheist then?'

'Look, I don't know about the label but my passion is that I don't believe that Living, magical, moment to moment Life is a Soulless, random, mineral coincidence and I'm disillusioned by the unreasonable materialist claims that it is and the use and abuse of science and biology, what is supposed to be the very science of Life, to try to make the point.'

'How do you mean?'

'I mean the whole point of Darwin's theory of evolution is that it is an attempt to provide an entirely materialistic, *'no-intelligence-needed-or-involved'* explanation of Life on Earth as a purely 'accidental' affair. Since Darwin published *On the Origin of Species,* in 1859, many have asserted that he succeeded in providing such an explanation. But, if you delve just a little, you discover that this is *not actually true.*

'You discover that far from Darwin's 'life-is-a-freak-fluke-that-evolved-by-accident' theory being proved true, *his ideas have never been confirmed,* certainly not by the fossils, that they have, in fact, been *contradicted* in numerous ways and from numerous angles – but that, so far, the mainstream biology community, which is predominantly materialist in its thinking, and incredible as it may sound regards Life and Mind simply as forms of 'accidental chemistry', refuses to admit this. I think this is an intellectual betrayal, Alisha. Things are not true just because a majority says so. They are true because they really are and the materialistic theories of evolution do not, despite vociferous Darwinian claims, fall into that category. Life clearly evolves over vast time but there is no evidence whatsoever that it does so 'by accident' any more than our cars, TVs and computers, all vastly simpler than any Life Forms ever arise and evolve 'by accident'.

'I have no idea why, let alone how, some kind of ultimate, 'super' or Higher Natural Consciousness or Supreme Intelligent Beingness – what many label God or Buddha Nature or the Tao or the Great Spirit and so on – becomes expressed as the millions of amazing, Soulful Life Form Beings on our planet – and, I suspect, on countless other planets throughout the Universe, but it is quite clear that the whole thing, in all its myriad facets, is an amazing display of 24/7– dynamic *intelligence, purpose* and *meaning* however mysterious that cosmic–scale intelligence and purpose may seem to us.

'We make incredibly clever things, like computers and TVs, using our *intelligence,* not by randomly tumbling their basic constituents in barrels. Yet, supposedly, that is what stunningly complex Living Cells and organisms, let alone consciousness and mind, are? The fluke products of the accidental concatenation of random atoms? Equally, where does our amazing *intelligence and motivation,* that enables us to perform so many astonishing feats, including brilliant science, come from? It is vastly in excess of any needs to gather food and purposelessly replicate genes in the meaningless Cosmos of the materialist imagination. The materialist thinking which dominates biology today holds that our very consciousness, our very sentiency, our very minds and intelligences are simply the freak, fluke effects of random events. Is this really true? There is no *evidence* for it.'

'In other words, you believe, Ollie, that the Universe itself is *meaningful and intelligent,* not mindless and accidental?'

'Yes, and it makes me sad that so many people today, based, more than anything, on the widespread belief in Darwin's never proven theory of evolution, believe that Life and existence are meaningless and random, that when we die that's it, and that all we ever were was a form of unnecessarily complex and accidental chemical fertilizer. You see, Ali, before Darwin, everything in Nature was *intelligent,* it had a *purpose, a Soul* and a meaning. Now, in much of mainstream science and biology, nothing in Nature is considered to be clever or to mean anything at all.

'Before Darwin the Living World was, to some extent at least, even in the west, considered to be magical and mysterious, even sacred, it was enchanted. After Darwin it became meaningless and accidental. Why I feel particularly strongly about this is that Darwin's theory was *never* genuinely proved true – the fossils never bore it out and as Darwin himself admitted the lack of confirming fossil evidence was, as he put it: "the *most obvious and gravest objection which can be urged against my theory*".[3] Rather, his theory was *agreed* to be true, as a more rational alternative to saying that the Living World was literally made by a kind of Father Christmas type entity, with a long white beard, out of spit and clay, in six very busy days, six thousand years ago.'

'Wasn't that a good thing?'

'Look, the fact that the Earth turned out to be billions of years old, not thousands, something India's yogis had worked out centuries ago by extra–sensory means, and that the Life Forms have endlessly come and gone, over millions of years, is *not 'proof'* that the whole thing has been some kind of prolonged *'accidental chemical reaction'!* Don't you see how absurd and unreal that is?

'I see what you mean. It does seem irrational to think of the Living World and our minds as the random and chance by products of a millions year long-running 'chemical accident' – or billions of them."

We walked on in silence for a few moments then I said,

'The problem with Darwinian–style, no-intelligence-needed theories of Life and Mind is that they do not conform to the data of reality. I don't believe in an anthropomorphic god with a beard at a clay modeling table in the sky somewhere but I feel frustrated with the materialist insistence, in biology today, that just because we can't *physically* see the Living World's subtle, inter–dimensional, 'higher' Natural causes that it can't have any and people are not to be allowed, on pain of damaging their careers in science, to *rationally infer* or intuit them, as all the giants of science prior to Darwin, and many since, always did. That doesn't make sense.'

---

[3] Darwin, C. 1859. *On the Origin of Species.* p. 292. Emphasis added.

'Why, specifically?'

'Because the fact that we cannot see something physically doesn't mean we cannot *rationally infer* or hypothesize its presence. On top of that, there have always been those with the developed psychic or spiritual gifts who are able, to some extent, by extra–sensory means, to attest to the real existence of Holistic Nature's intelligently causal multi–dimensions – India's yogi–adepts, for example. You see, we have two primary ways of obtaining information or raw data about existence. Firstly, through our physical senses and secondly, through a collection of inner or higher senses which we all have but which are highly developed in some and hardly at all in others.'

'What kinds of senses do you mean?'

'I mean our faculties of abstract thinking, imagination, intuition, inspiration, psychic knowing, telepathy, remote viewing, shamanic vision, spiritual revelations and so on. We are, most of us, reasonably good at thinking, with some imagination and intuition, and can come up with ideas of various kinds. A smaller proportion of us are born with some of the more rarified of these abilities, such as remote viewing, the capacity to see into other times and dimensions[4] and into the deep nature of matter itself,[5] and the rest of us can develop such higher senses and abilities to a greater or lesser degree, through various forms of inner, psychic and spiritual training, like the spiritual–scientist, yogi–adepts of India who, centuries ago, using such inner means of higher–sensory perception, had seen into the past of the Universe and worked out that it was billions of years old, not thousands, as was believed in the west until the nineteenth century.

'By these same means the yogis had also worked out that Life on earth underwent vast cycles that they called yugas in which the Living World experienced times of expansion and Life Force renewal and periods of decline, which may, in fact, correspond to the patterns of renewal and decline so clearly seen in the fossil records, the pattern of evolution that some have called 'punctuated equilibrium'.[6] The abrupt, geologically speaking, appearances of new species (what the late S J Gould called punctuations), their stable presence over millions of years (equilibrium) followed by their relatively rapid disappearances (extinctions) with no signs of the cumulative and gradualistic evolution of the Life Forms that Darwin's theory predicted should be taking place during the long

---

[4] Yogananda, P. 1946. *Autobiography of a Yogi*. Self-Realization Fellowship.
[5] Tompkins, P. 1997. *The Secret Life of Nature*, Thorsons, pp. 68–69.
[6] Davidson, J. 1992. *Natural Creation or Natural Selection*. Element Books.

periods of stasis or equilibrium. The Yogis also observed, by these same inner, higher–sensory means, that 'matter' itself was not quite as it first appears. That it was, in some senses, a vast illusion made of nothing but 's—p—a—c—e', of 'mind–stuff', of consciousness, purpose and intelligence that the yogis called 'chitta'. An insight now confirmed by the discoveries of modern quantum physics.[7]

'Even some scientists in the physical–science tradition of the west have used intuition and imagination, that is 'going within', to come to their conclusions about the workings of outer Nature. Einstein, famously, imagined himself riding on a beam of light when he was formulating his theory of relativity.

'The point is, we do not obtain all our information about reality purely through our physical senses. In fact, we have various other senses and just as our physical senses, and their mechanical extensions, pick up the data of the dimensions and vibrations of physical reality so can our other senses, to whatever extent they are developed in us, help us to tune into other dimensions, vibrations and frequencies of existence.

'Modern, materialistic science, however, tends to downplay the role of imagination, intuition and inspiration in scientific discoveries and it believes that all the information that scientists work with comes solely from their physical senses and the analysis of purely physical information. This is because materialist science tends to dismiss, in spite of the large body of modern evidence for it, the existence of higher-sensory or extra-sensory perception.[8] It also denies that we are multi–dimensional beings, that Life after death is real[9] and that it is possible to communicate telepathically with those who have passed on.[10]

'But whether we gain information through the physical senses alone or higher–sensory perceptions as well, that is just data gathering. It is after we have gathered the data that we can bring our intellects, privately and collectively, to bear on the information

---

[7] Capra, F. 1992. *The Tao of Physics*. Flamingo. Goswami, A. 1993. *The Self-Aware Universe: How Consciousness Creates the Material World*. Jeremy P. Tarcher.

[8] Radin, D. 2009. *The Conscious Universe*. Harper One. Carter, C. 2012. *Science and Psychic Phenomena: The Fall of the House of Skeptics*. Inner Traditions. McLuhan, R. 2010. *Randi's Prize: What Sceptics Say About the Paranormal, Why They Are Wrong, and Why It Matters*. Matador.

[9] Fontana, D. 2005. *Is There an Afterlife?: A Comprehensive Overview of the Evidence*. O Books. Carter, C. 2010. *Science and the Near Death Experience*. Inner Traditions.

[10] Schwartz, G. E. Phd. & W. L. Simon. 2002. *The Afterlife Experiments: Breakthrough Scientific Evidence of Life After Death*. Atria Books.

and use it to help us to understand reality, how to proceed and how best to live.

'Humanity's spiritual and religious paths are all based, in part, on this kind of *inner* data gathering, the data of intuition and higher-sensory perceptions, visions and revelations. According to the eastern spiritual traditions, Taoism, Hinduism and Buddhism, one of the goals of humanity's inner development is to relate to the Mystery of Existence in such a way that, ultimately, subject and object become united in understanding and perception. The benefit of this is that it transforms and heals the painful, 'separative', dualistic perception of Reality that humanity has fallen into. The process of enlightenment leads to the experiential realization of our essential oneness with all that is, a realization which brings profound inner changes in its wake – joy, inner freedom, increased vitality, personal capacities and peace.

'According to these traditions, this is only possible because we are integrally part of a meaningful, multi-dimensional Cosmos that is made not only of clever 'outer' mechanisms, such as gravity, light, electricity, the smart–materials–chemicals and so on but also of consciousness, intelligence, love and purpose. This means that, as conscious, self-aware beings of mind and Soul, we are able to use that very consciousness and intelligent awareness to explore the multi-dimensional interiority of the intelligent matrix of consciousness and meaning from which we emerge and, ultimately, by transcending and dissolving the illusion of separation and separativeness to reunite with it in a deeply fulfilling way – what the eastern traditions call self-realization or enlightenment, and the yogi–adepts, of the Hindu tradition refer to as jivanmukta or liberation in this Life.

'On the other hand, the separative, reductive, mechanistic, western way of materialist science dismisses the idea that we can obtain any real knowledge of Life and existence by such inner, intuitive and higher–sensory means and, as it observes outer Nature, far from seeing the ultimate Oneness of All that Is, sees separation, unintelligence and 'accident' wherever it looks.

'It then seeks to relate to what it sees as the *unintelligent* and random mystery of existence by pursuing *intelligent* science to explain how *meaningless and random* it all is – reducing all of Nature to mindless and accidental machinery, from gravity, light and electricity to accidental atoms and molecules to accidental cells and plants to accidental animals, people, planets, stars and galaxies.

'Between these two ways of knowing and perceiving, the spiritual one that obtains its data about total Holistic Reality partly by inner means and the materialist one that focuses exclusively on what it regards as 'accidental' mechanical externalities, seeing Soulless, ghost-less 'machines' all the way up and all the way down – and no true Mind or Soul or Subjectivity or Interiority anywhere at all – there's a third way, scientific or philosophical Idealism.

'What's philosophical Idealism?'

'It's the classical 'mind-before-mechanism' approach to reality. Unlike the interior focus of the religious and spiritual paths, and their close cousins the artistic and poetic ways, philosophical Idealism is, primarily, an exterior path of knowledge but, by contrast with philosophical materialism, it is one that sees the outer Universe as a multi-dimensional, multi-vibrational emanation of consciousness, intelligence and purpose.

'This is because, philosophical Idealism, as it studies the world, notices the intelligence, the astonishing 24/7–dynamic functionality of Nature. On the otherhand the materialist approach maintains that Nature may 'look' clever but really what we are using our own keen *intelligences* to observe are the second by second effects of mindlessness, *unintelligence* and accident at work.

'Yet, if something is 'working' or 'functioning' it has a purpose or teleology. That's just a fact, not a matter of opinion. Idealism notices, Ali, that the properties of Carbon and all the things it can do are incredibly smart. It doesn't, however, attribute Carbon's intelligent and purposeful properties to any kind of medieval bearded–god type creative agency. It has no need to. This is because, by contrast with the materialist view, it realizes that the non-existence of 'bearded-gods' doesn't prove that total Nature is unintelligent and accidental. It realizes, along with Aristotle, long ago, that regardless of why Carbon exists – and saying it's because of the Big Bang does not take away from this – that because it 'does' things, that because it has 'functions' it really *does* have a purpose or many purposes – what some people call teleology.

'If we deny Nature's obvious purposefulness and intelligence we make our language meaningless and dishonest. Unfortunately, materialism, necessarily, makes both our language and our science false to reality. This is because if clever Carbon, Hydrogen and Oxygen, along with all the other smart–materials–chemicals, are, as materialism insists, 'purposeless' and 'accidental' then everything is, all the way up and all the way down, including anything anyone says, scientists and non-scientists alike.

'When idealist thinking science studies digital DNA coding, "like a computer program but more advanced than any software ever created", as Bill Gates put it, it notes that modern probability and information theory predict that no coding system can ever arise 'by accident', it's not merely improbable, it's impossible,[11] regardless of the coding system's unknown origins – a conclusion the logical implications of which materialist thinking science firmly resists or simply ignores. The idealist view doesn't pretend to know the final interdimensional source or sources of the DNA CODE but, unlike the materialist position, it doesn't believe that 'accidents of chemistry' can create CODES.

'One of the great positives of the idealist or non-materialist way of looking at Nature is that it re–intelligences and re–purposes the Living World. It points towards the intrinsic meaningfulness of existence. The idealist view holds that when doing science we don't have to deny all the obvious intelligence, purpose and Soul in Nature just because the intelligently causal elements of existence, whatever their precise nature, cannot be physically seen but can only be rationally inferred or intuited or because Nature can seem so 'weird'.

'Weird' doesn't imply unintelligent. An intelligently 'weird' universe, that contains qualities of consciousness, sentiency and purpose, a vast Living, breathing, multi-dimensional organicity, corresponds much better with the data of Reality than the materialist hypothesis that we live in a mono-dimensional, mindless and unintelligently 'weird' and mysterious universe. Existence may be strange, surreal even but is it 'mindless and accidental' as materialist science believes?

'The fact that there are no 'bearded–gods' anywhere to be seen making and sustaining everything, second by second, 24/7, is hardly extraordinary evidence for such a claim. Pythagoras, Socrates, Plato, Aristotle, Buddha, Lao Tzu, Jesus and Mohammed were all well aware that Life's intelligent sources did not assume such forms, and that they could not be physically seen, but could only be rationally inferred, intuited or, to some extent, directly experienced by psychic and spiritual senses.

'Paul Davies writes:

> 'The living cell is **the most complex system known to man**. Its host of specialized molecules, many found nowhere else but within living material, are themselves enormously complex. They execute a dance of exquisite fidelity, orchestrated with breathtak-

---

[11] Gitt, W. 2006. *In the Beginning Was Information*. Master Books, p. 124.

ing precision. ... the dance of life encompasses countless molecular performers in synergetic coordination.'[12]

'In materialist biology this purposeful, intelligent, Life–Force driven bio-functioning is believed to be an 'accident'. This is unreasonable.

'Is the functioning of a computer *unintelligent or 'accidental'?* Did no *intelligence and wisdom* go into its making? Is the functioning of even just one tiny Living Cell, "the most complex system known to man", vastly more complex than any machinery we make, *unintelligent and accidental?* The methodologically materialist thinking, that governs biology today, replies 'yes' to that question.

'The exciting thing is, Ali, that, today, there are scientists and thinkers who are beginning to show empirically, scientifically and mathematically that it's *impossible* to genuinely explain Living Nature and our own minds and intelligences as the 'random' and unintended products of *unintelligent* processes, no matter how fond some people seem to be of the mindless and meaningless Cosmos of materialist faith and belief. It is their discoveries, happily, that are helping to restore the *Intelligence, Soul,* and meaning back to *Life* and Living Nature that Darwinism's non-super Naturalist or anti–spiritual theory and its nearly wholesale adoption by the scientific community sucked from them.

'Darwin's materialist theory of evolution de-Souled and disenchanted the Living World. It sought to kill poor Tinker Bell. But was he really right? Or was the adoption of his theory just a way for a new kind of materialist, pseudo high priesthood to take over from the Anglican brahmins and divines of Oxford and Cambridge who Darwin and Huxley overthrew in their stunningly successful intellectual coup d'etat as the arbiters of the highest and most important truths at our culture's intellectual top tables?

'Richard Dawkins says on page one of his book, *The Blind Watchmaker,* that the amazing Life Forms *'appear to be designed for a purpose'* – but, of course, he believes, are not. I believe, however, that *the reason* that the Living World looks, as even Dawkins acknowledges, *'designed for a purpose',* is because *it really, genuinely is.* It's appearance of intelligence and purpose is not an illusion, it's real.'

'Ollie, personally, I agree with you. But, where's your evidence? I want the Living World to be meaningful and intelligent, rather than a random cosmic 'accident', but surely all these important scientists and

---

[12] Davies, P. 1998. *The Origin of Life*. Penguin. Emphasis added.

thinkers like R. Dawkins can't all be wrong? Perhaps if they all believe that the Living World is simply a very long-running *'accidental chemical reaction'* that never stopped for billions of years, then it really is.'

'OK, Ali, humble bacteria. They have been here since the very beginning of time, around 3.8 billion years ago. There is *no evidence anywhere* that they gradually evolved, Darwinian style, through processes of slow, gradual, ever rising complexity into being 'accidental' bacteria. And, these so-called simple, primordial bacteria, here, right from the beginning, all perfect and complete, 3.8 to 3.5 billion years ago, are, you may be surprised to learn, not simple at all! They are *stunningly* complex mechanisms! They are far more cleverly complex than *anything* we make, tiny as they are. They are literally, each and every one, like 24/7 Living, self-reproducing, micro–miniaturized chemical factories of breathtaking complexity. Living Cells are, as Paul Davies says, "the most complex system known to man"! It is scientifically unreasonable to claim that they are the astonishingly clever results of 'purposeless', Life–Forceless, zombie chance, of 'accidental chemistry' just because we cannot directly or physically see Life's subtle, super–intelligent, higher Natural laws, forces and processes operating second by second, 24/7.

'As Carl Sagan said "extraordinary claims require extraordinary evidence". What is more extraordinary than the claim that 'dead', random chemistry leads to the Living World? It's irrational. There's nothing in the laws of 'accidental', Life–Forceless chemistry to entitle anyone to make such a claim, let alone any 'extraordinary evidence' to back it up.'

'You mean we've no reason to believe this unless we are of the materialist faith and belief that existence itself is meaningless and accidental – that there are no *intelligently causal* elements to reality? No unseen multi–dimensional laws, forces or agencies to which people give blanket names like God, Buddha Nature, the spiritual dimensions or 'super', Higher Nature?'

'Yes. That *is* the *only* reason to believe that Life on Earth is a billions-year-long running *'accidental chemical reaction'* that just never stopped.

'When you put it that way it sounds ridiculous.'

'I know, Ali, of course it does, but that is the logic of the materialist position.

'The trouble is that right now the materialist idea that Life is just a weird and meaningless 'accident' *is* the scientific-consensus view – a kind of modern equivalent to the Ptolemaic or pre-Galileo position that used to afflict astronomy. It doesn't fit the data of reality, it entirely conflicts with humanity's perennial intuitions and spiritual revelations that Living Nature is intelligent and meaningful, there is *no genuine evidence* for it, but a lot of people believe it because some scientists 'say so'. This is about as rational as believing that the world was made out of spit and clay by a bearded–god in six busy days six thousand years ago because some biblical literalists 'say so'. It is less rational, in fact, because the world's creation stories were only ever intended as metaphors. They simply embodied the universal insight that holistic and multi-dimensional Nature has blazingly intelligent sources and that intelligent mechanisms only ever have intelligent causes not mindless accidental ones.

'It is materialism that has made our science and our predominant world–conception so unreasonable by making the anti-empirical claim that 'chemical accidents' are supremely creative and evolutionary. However, in scientific–consensus reality what is 'true' is not necessarily what is actually true but, predominantly, what the majority of scientists say is true, just as even *after* Galileo it continued to be 'true', for a while, that the Sun orbited the Earth. So for now we are saddled with the unreasonable contradictions of materialist science and biology: 'accidents build', 'mindlessness leads to minds', 'unintelligence leads to intelligence' and so on.'

Ali smiled and looked at me quizzically. I said:

'Remember, microbial Life has been here from the very beginning, 3.8 billion years ago, with no evidence whatsoever that it arose by Darwinian style accident. Listen to what Michael Denton, a molecular–biologist says about this:

> 'Although the tiniest bacterial cells are incredibly small ... each is in effect a veritable micro-miniaturized factory containing *thousands of exquisitely designed* pieces of intricate molecular *machinery,* made up altogether of one hundred thousand million atoms, *far more complicated than any machinery built by man* and absolutely without parallel in the non-living world.'[13]

'Alisha, is it really reasonable, when beholding such astonishing, 24/7–Living complexity, for our scientific culture to continue to insist, in a kind of Naked Emperor in reverse way, that Life and the Living World 'must be' unintelligent and 'accidental' phenomena, when we are trying to make sense of them?'

---

[13] Denton, M. 1985. *Evolution: A Theory in Crisis*. Burnet Books, p. 250. Emphasis added.

'The problem with this materialist view of Life and evolution is that (a) we have to use our mysterious, immaterial Soul quality of *intelligence* to argue that the Living World may *'look'* incredibly clever, stunningly so but *actually* we are all deceived – really it's just the amazingly clever 'seeming' (only) result of utterly unintelligent and chance processes.

'If that were actually true where do the minds and *intelligence* that allow us to conclude that nature is supposedly so very *unintelligent* come from? From outside the *'mindless and accidental'* universe of the materialist faith system, perhaps? (b) Materialism also causes us to insist that collective humanity's *nearly universal* intuitions that there *really is* more to Life and the Living World than meets the immediate physical eye, that there really *are* deeper meanings and *higher* purposes to things, are all just so much make believe and wishful thinking.

'What I find so disappointing about that attitude, Ali, is that it's not curious and genuinely inquiring as science was supposed to be. It's not open-minded, let alone scientifically curious of the countless psychic, religious, mystical and spiritual experiences that so many people have had[14] – not just the great saints and sages and founders of the world's religious and spiritual traditions but also many perfectly *ordinary* people (and possibly even *most* people) have *at least one or two* mystical or peak experiences, in their lives, that point to these deeper and higher meanings. The materialist attitude to all such experiences, however, is that if you can't weigh them, measure them or count them in a purely material way then they don't count as part of the meaningful data of reality or have any true knowledge value at all.

'Although materialist thought has no way of proving its claim that they are not useful data, gathered by intuitive or extrasensory means and that they have no knowledge-value as to the true and multidimensional Nature of the totality, that is its belief. As to the creative or evolutionary powers of 'accidental chemistry', the genuine evidence is that, at the most, it can create pretty crystals which some people believe could, somehow, accidentally become Living, breathing, thinking, feeling, creating, Soul–Life–Form Beings. This,

---

[14] James, W. 1902. *The Varieties of Religious Experience*. Penguin Classics. Bucke, R. M. 1905. *Cosmic Consciousness: A Study in the Evolution of the Human Mind*. Innes and Sons. Kelly, E. F. & E. W. Kelly, A. Crabtree, A. Gauld 2009 *Irreducible Mind: Toward a Psychology for the 21st Century*. Carter, C. 2012. *Science and Psychic Phenomena: The Fall of the House of Skeptics*. Inner Traditions. McLuhan, R. 2010.

however, doesn't work. As origin of life researcher, Leslie Orgel pointed out:

> 'Living things are distinguished by their ***specified complexity***. Crystals such as granite fail to qualify as living because they lack complexity [they are specific but not complex]; mixtures of random polymers fail to qualify because they lack specificity [they are highly complex but randomly so, meaninglessly so].'[15]

'But, that, in any case, assumes, Ollie, that the processes that cause even simple crystal formation are, themselves, *unintelligent* and 'accidental'. I don't see the logic in that. If the universe as a whole is intelligent then its laws, including those that govern crystal formation, will also be intelligent laws and processes.'

'Yes, I agree, and the point is that *there is no evidence* that accidental chemistry, including accidental crystal chemistry, can create *working 'machines' of any* kind let alone the most clever ones known to humanity – the *Living, Breathing, Breeding,* Life–Form–machines, if we must use mechanistic language. As Michael Denton says:

> [The] break between the living and nonliving world… [is] the *most dramatic* and fundamental of all the discontinuities of nature. *Between a living cell* and the lowest highly ordered *nonbiological* system, such as a crystal or snowflake, there is *a chasm* as vast and *absolute* as it is possible to conceive.[16]

'There's *nothing* in the laws of 'dead' mineral chemistry, including crystal chemistry, unaccompanied by subtle Life Forces and Soul Forces or other super-ordinate, 'higher' Natural laws and processes (whose rational existence materialist thought refuses to acknowledge, intuit or infer), that gives us any reason to believe that the Living World continuously arises, *right now, 24/7, 7/52* by Life-Forceless and accidental, 'zombie' processes.

'It is only if our foundational scientific hypothesis and belief is that very existence itself, the entire universe is unintelligent and random that we *have to* assume as methodologically naturalist or

---

[15] Orgel, L. 1973. *The Origins of Life*. John Wiley; p. 189. Emphasis and text in square brackets added.

[16] Denton, M. 1985. *Evolution: A Theory in Crisis*. Burnett Books, p. 249, emphasis added.

methodologically materialist science does, that no matter how clever and purposeful Life and the Living World 'look' that really they are 'no-intelligence-needed' phenomena. But this is a philosophical impossibility. We cannot *logically* argue that Living Nature is a purposeless and *unintelligent* phenomenon without having to use our own *very intelligence* to argue that *we must be unintelligent too!* It doesn't work. It's like the statement, 'I never tell the truth.' It doesn't make sense.

'The modern, materialist dogma that *accidental, 'dead' chemistry* created and *is, right now, ongoing mindless second by meaningless second,* what the dynamic, Living, Breathing Soul Life Form Beings *are* is not based on weighed and measured and counted evidence, Ali. Rather it is a non-negotiable materialist faith position, no less unquestionable, in its own unreasonable way, than some dogmatic religious doctrines.'

'At root this dogma is based on little more than the Victorian realization that (a) the world's scriptures were of no knowledge value in a *physical–science* sense, even if they contained a lot of truth in a *spiritual–science* or religious sense, (b) that there was no naive, bearded-god type Creator Being anywhere to be seen, even in the best Victorian telescopes and (c) the fact that someone artificially synthesized Urea in 1828, which was taken to prove that there were no superordinate Life Forces and probably no Soul either.

'None of these points, however, are good enough evidence for the materialist faith that Living Nature is a billions-year-long running *accidental chemical reaction.*

'Look, the truth of the knowledge claim that Mr. Rolls and Mr. Royce's magnificent Cars had an intelligent genesis and evolution through time is not falsified by the naive and unscientific Creation Story of the Rolls-Royce that goes:

> 'Once upon a time, Messrs. Rolls and Royce said, *"Let there be magnificent, shiny Rolls-Royce Cars!"* and lo, there were, and it was good. And they said to the people, please look after our magnificent, shiny Cars and Love them and treat them well, etc.'

'Is that creation story not perfectly true, in its own simplistic way?'
'I suppose it is true in a naive or poetic sense.'
'Exactly, so while this poetic explanation of the elegant Cars' genesis is not a mechanical engineer's description of the process, it is still perfectly true. It is, more or less, even literally true. Equally, the world's

creation stories, while of no use in a physical–science sense, truthfully told humanity and continue to tell us that we live within an *intelligent* and *meaningful* total Reality, not an unintelligent and random one. Those stories are not falsified because they are not astrophysics or biochemistry text books.'

2 Idealism: Life is Intelligent – The Classical View

Idealism is the classical 'mind-and-intelligence-before-matter-and-mechanism' view of existence. Existence is like a two-sided coin, a two-sided unity, consisting of inner and outer, subject and object, spiritual and material, Life and Form, Soul and mechanism at all levels and dimensions.

'The eminent physicist, Sir James Jeans was an idealist. He said he inclined to:

'the idealistic theory that *consciousness is fundamental,* ... the material universe is derivative from consciousness, not consciousness from the material universe...'[17]

'Antony Peake says:

'Self-awareness within a seemingly unaware universe simply does not make sense ... [science may be coming to] a model in which mind creates matter rather than matter creates mind.'[18]

'So you believe, Ollie, that we cannot explain evolution as a purely materialistic process, in the way we *can* explain this randomly evolved river valley in terms of 'just chemicals plus accidents' (earth, rocks, water, gravity, etc) but only in terms of chemicals in combination with intelligently causal sources?'

'Yes. The materialist, ('accidents-create-clever-mechanisms'), view of the Great Mystery of existence is that matter, that was itself an 'accident', by a long series of entirely *unintelligent and chance* processes eventually became the amazing Living World with its fabulously clever bio-machines (if we must be mechanistic) and *incredibly intelligent* human minds (pervading those bio-machines) including those of Buddha, Plato, Newton, Bach and Einstein.'

'The materialist view of the Great Mystery of existence is that random, 'mindless' matter which ***just exists*** (who knows *why?*) leads

---

[17] *The Observer* – http://en.wikipedia.org/wiki/James_Hopwood_Jeans
[18] Peake, A. *The Out of Body Experience: The History and Science of Astral Travel.* Watkins Publishing. Text in square brackets added.

to accidental *intelligence* and accidental *minds* by *mindless* evolution over vast time. The extraordinary claim – with no extraordinary evidence ever found to back it up – is that just Lots of Time + Lots of Chemicals + Lots of Mindlessness + Lotts of Unintelligence + Lots of Entropy + Lots of Accidents eventually led to the fabulously complex Soul–Life–Form Being mechanisms that are the *Living, Breathing, Breeding* Plants, Animals and People, including our own amazing human minds and intelligences.

'The classical, Idealist view of the Great Mystery of existence, on the other hand, is that astonishingly clever, cosmically scaled Consciousness, Mind, Purpose, Desire and Intelligence, which also *just exists* (who knows *why?*), plus the amazing smart-materials-chemicals matter, that It gives rises to, hologrammatically Self-fragments into trillions of lesser intelligences and subjectivities, working synergistically in many nested hierarchies of intelligent activity leading to our Living World – and probably many others across the universe.

'Materialism disagrees. It says the Cosmos is 'mindless, unintelligent and accidental'. It believes that the incredibly precise and non-random laws of physics are meaningless coincidences, random effects of some kind of multiverse where because, it is imagined, accidental universes are arising all the time, in fact trillions might have arisen even while we've been speaking, some are bound to 'seem' intelligent even though, really, they are not.

This is like saying Einstein only 'seemed' to be intelligent but he wasn't really. $E=mc^2$ only 'seems' to be intelligent but it isn't really. This materialist way of thinking de-intelligences Nature and makes the universe we are actually in meaningless. It means that our Star system with its beautiful, mysterious orbiting Planets 'has to be' an accident, that the intelligent functioning of the continents, oceans, mountains, glaciers, rivers and lakes of our beautiful bio-dome, desert island paradise planet, in the shining ocean of space, Soulful, magical, amazing, mysteriously self-regulating Gaia, also all 'have to be' flukes and that our very *consciousnesses* and *intelligent* minds also 'have to be' accidents.

'Yet, even since Darwin, prior to whom materialist thinking was always a minor strand in western thought, there have always been idealist thinkers who disagreed with the pessimistic, materialist conception of Life and existence as, ultimately, meaningless and random. Philosophical Idealism sees Consciousness, Awareness, Intelligence, Potentiality, Potency and Purpose as *primary conditions* that form the fabric of existence, including the *maya* of apparently solid matter 'made', as the yogis described centuries ago, of Consciousness and Intelligence, that is of nothing but intelligently informed quantum-energy S—P—A—C—E and that it is these *same* qualities that make even apparently 'mindless' atomic m—a—t—t—e—r itself so clever.

'Think of all the smart-materials-chemicals that make outer, 'mechanical', 'material' Life and the Living World possible. Carbon, for example: it's an amazingly versatile atomic mechanism, without which our chemical bodies would not exist, nor, indeed, many of the things we ourselves use it to make. The Idealist understanding is that we can no more take the ontological *cleverness and teleology* out of smart-materials like Oxygen, Carbon, Hydrogen or Nitrogen, which provide the main components of the complex chemical compounds that go, externally, to build the Life Forms, than we can take the warmth out of heat.

'The materialist reply to this is that amazingly versatile, smart-materials Carbon is just another accident or meaningless coincidence of the multiverse. Ideological materialism in science 'makes meaning' by denying meaning, it uses mind to deny mind, it uses intelligence to deny intelligence, it uses the purposeful activity of science to deny that there is any purpose or teleology in Nature at all. And it's extraordinary evidence for these extraordinary claims?

> 'The world's scriptures are not biochemistry textbooks and we can't see a 'bearded–god' anywhere, not even in the best telescopes.' 'In materialist thinking, Carbon has no meaning, *intelligence* or teleology, it's just functionlessly and accidentally there, along with all the other billions of 'accidents', out of which materialism insists Nature is composed.'

'Ollie, that doesn't make sense. If something has a function, like Carbon, that helps to build all the Life Forms then it *does* have a purpose. Logically, it simply cannot be lacking in teleology.'

'I agree. *Everything in total Holistic Nature has a function or purpose* but that functionality or teleology is especially apparent in the Living World. This is why I'm passionate about this subject because the materialist thinking that dominates biology today doesn't make sense.

'It insists that All that Exists is nested hierarchies of pointlessly and mindlessly 24/7 functioning 'accidents', all the way up and all the way down. What could be more counterfactual and unreasonable? Ideological materialism in science is like a Soul corroding, universal acid of untruth that seeps everywhere making Life meaningless and cutting us off from all knowledge of the highest and most precious things, the deeper and higher Sources of the amazing multi–dimensional reality in which we find ourselves as Beings of both Space and Time and Beyond Space and Time.'

'What do you mean by total Holistic Nature?'

'I mean, firstly, all that exists – that is all the dimensions and vibrations of existence, physical and non-physical, within Space–Time and Beyond Space–Time. Secondly, I mean the idea that total Nature while a Unity or non-dual is, paradoxically, at the same time dual; like a coin it's two sided. One side, 'tails', describes 'ordinary', visible, objective, chemical, Nature within Space–Time, of which our bodies, these grassy meadows, these trees are all parts.

'The other side, 'heads', describes the fifth and higher dimensions, beyond Space–Time, subtle, *intelligent*, Spiritual or 'super' Nature of which our universal Witnessing Awareness, our Subjectivity, our Intelligences, Minds and Souls are all parts. Our *Subjective,* feeling, thinking, desiring, loving, creating and willing belong to the 'super' Natural or spiritual side of the coin of Reality which, ultimately, is beyond Space–Time. Our *Objective*, chemical bodies are but our outer garments, our physical homes during our physical lives, within Space–Time, of our 'super' Natural witnessing awareness, our subjectivity, our intelligent, Minds and Souls.'

'So, according to the Idealist view of reality total Holistic Nature has both *subjective,* inner dimensions or 'heads' sides and *objective* outer dimensions or 'tails' sides?'

'Exactly. And the Idealist view of existence is that it is the Subjective, creative, *intelligently* causal *multi-dimensions* that are the ultimate *intelligent* (not random) evolutionary sources and causes of our challenging, beautiful, strange and magical World.

'Materialist–monist thinking, on the other hand, sees total Natural Reality or 'All that Is' as mono-dimensional, as a purely *one-sided* coin, all 'tails' and no 'heads', fundamentally all object and no true subject or mind or intelligence anywhere to be seen – all accidental chemical machine and no true mind, Soul or ghost *in* the machine.'

'But machines never arise by accident do they? And what about us and these animals in these fields? We are all sentient Beings, with witnessing awareness, subjectivity, sophisticated or primitive minds, intelligences, wills and purposes of various kinds. Surely that does mean that there is a 'heads' side or Subjective, ghost in the machine dimension to total Nature?'

'I agree, Ali, that is the way the Idealists see it. But the materialist view is simply that there *is* no true 'heads' or Subjective side, there *is* no Self-Existing, Self-Arising, Cosmic-Witnessing Consciousness that creates and collapses the universal quantum wave function and brings

All into Existence. Materialist thought believes that there *is* no total Holistic Nature, that is both subjective and objective but rather that there are just objective, chance chemicals (generated by a random and meaningless Big Bang) + 'accidental' chemical laws + a plethora of chemical accidents and that's what Life, Mind, Intelligence and the Living World also are – 'accidents' of random atoms of carbon etc.

The trouble is, materialism in science and biology is not an empirically-based view of reality. No one has any good reason to imagine that dead, Life–Forceless, *accidental chemistry* creates anything clever – let alone that it would do so continuously, second by second over millions of years. The materialist dogma that controls biology today is that lightning struck ancient seas and "Life" (whatever that might truly be) randomly started and, eventually, solid, stolid, insentient little billiard-ball atoms of (allegedly) accidental and 'purposeless' carbon, oxygen, hydrogen and nitrogen accidentally became the Sentient, Intelligent, Subjective, Creative, Thinking, Feeling Beings, of the Plants, Animals and the People, including the Minds of Socrates, Plato, Galileo, Beethoven and Einstein. This is as rational as saying:

> 'Volcanoes and Earthquakes took place which 'accidentally' led to copper and iron ores and crude oils being turned into wires, circuit boards and transistors, then to 'accidental' TVs which then began 'accidentally' making TV Shows.'

'Our implacable, real-world experience is that intelligent mechanisms and 'machines' *always* have intelligent causes. Only by turning logic and all our experience upside down, by arguing for what is *a vast distortion or, at best, a peculiar delusion,* that the Life Forms' breathtakingly smart, Life–Force driven, 24/7–clever bio-functionalities, compared to which all our evolutionary creations, cars, computers, TVs etc, are positively simple-minded, are 'not really' clever at all, they're not intelligent or smart, they only 'appear' to be, can one then start to build the counterfactual materialist argument, the strange delusion or collusion that blue is green and green is blue, and that the most *intelligent* and clever mechanisms we know of, the Life Forms, *actually* arose and continuously arise right now, second by second, 24/7, 7/52 by utterly mindless and accidental processes – amino acids bumped into each other and there you have it, tiny, fabulously complex, digitally DNA CODED Living Cells, more complex than Jumbo Jets and Space Shuttles combined.

'This is as reasonable as arguing that a Jumbo Jet or Space Shuttle, is an *unintelligent mechanism,* there's nothing clever about it, *no intelligence* or teleology went into it's shaping or making. However, by adopting this position materialist science and biology can then argue:

> "Well its ultimate or final sources and causes can't be *intelligent or purposeful* either can they? The Jumbo Jet or Space Shuttle or Living Cells (far more complex than Jumbo Jets and Space Shuttles combined), 'must have' just arisen by accident then."

'This is what the depressing materialist ideology of 'only-matter-is-real-and-Life-is-an-accident' has done to our science and culture, Ali. It has made them unreasonable and false to reality.

'The idealists, on the other hand, say that our minds and intelligences, and, indeed, those of all Living and Sentient Beings, flow from the *essence* of a Holistic, multi-dimensional Nature that is, itself, a vast, dynamic, creating, evolving, expressing, unfolding, extending Consciousness and Intelligence. Their view is that we are among the offspring of this *Living Intelligence* – that some call Supreme Being, God, Brahman, Allah, Buddha Nature, the Tao etc – which itself is made of Intelligence, Love and Purpose, it is the very essence of these immaterial, Soul and Spiritual qualities. We are the children of intelligence, purpose and meaning, not of 'mindlessness and accident'. It is this very fact that means that we ourselves, brilliant, at times, microcosms of the macrocosm, in some senses 'made in these images', are able to create so much magic and meaning – think of Buddha, Jesus and Muhammed, Shakespeare, Mozart and Beethoven, Plato, Galileo, Curie and Einstein ...'

'I'm slightly losing you here, Ollie ...'

'What I'm trying to say is that the classical, 'mind-before-matter-and-mechanism' view of Nature is that it is made of and is an expression of, among other things, the very *essence* that we call *intelligence*. It's very nature is a self-arising, self-existing *intelligence, purpose and wisdom* and that this *essence of purposeful intelligence* pervades and constitutes the very fabric of space–time–energy–and–matter at all levels, dimensions, vibrations and frequencies of existence. This is *why,* for example, the smart-materials-chemicals, like Oxygen, Carbon, Hydrogen, etc, are so clever and have such amazing functionalities or teleologies, because they are all part of the intelligence and teleology which forms the very fabric and deep structure of total Reality.

'This is *why* Nature is both *intelligent* and so amazingly *intelligible*. This why we are *intelligent* and why we can do *intelligent* science. This is why Einstein not only 'looked' intelligent but he was intelligent. This is *why* E = mc2. This is *why* atoms and molecules and gravity and Living Cells and plants and animals and people and planets and stars and galaxies all *work* in stunningly–smart, 24/7–dynamic nested-hierarchies of synergistic and holistic functioning, because *the whole thing,* from top to bottom, is an expression of *purposeful intelligence* and intelligent Life and Soul-Force driven activity.

'Materialist thought, on the other hand, doesn't see Holistic Nature as *intelligent* or purposeful at all – quite the reverse. Wherever it casts its gloomy, skeptical gaze in Nature it sees meaninglessness and fluke. Everything is supposedly an 'accident', and there's no rhyme or reason or justice or fairness or love or intelligence or meaning or purpose to any of it.'

'People who think like that have a point there, Ollie, don't they?'

'Yes and no. They are right to notice how harsh and cruel this world can be, not least Living Nature, but they are making the mistake of confusing the moral qualities of something with its causation. Many people would argue that military weapons, like attack helicopters or stealth bombers are of dubious moral value. Their value is ambiguous at best, perhaps good for protecting friends but terrible to others. However, no one doubts the lethal *intelligence* or purposefulness that they embody.

'We should no more doubt the obvious *intelligence* and *purposefulness* or teleology of the countless biological mechanisms of the Living World, including various creepy crawlies, for example, either just because we don't like many of them, find them baffling or cannot understand their reason for being. Why nettles and thistles, newts, tortoises, butterflies, giant sequoia and jacaranda? It doesn't, however, do our own, inborn *essence* of *intelligence* any credit to insist that these strange, magical Soul–Life–Form Beings, be they nettles, brambles and bindweed or wasps, poisonous snakes and spiders, are ongoing *Darwinian–accidents,* just because we find them weird! Weird does not equal unintelligent or random.'

'Doesn't that view make the Living World's *intelligent,* spiritual or multi–dimensional sources and causes pretty hard to understand then?'

'Look, it may do. I'm not trying to make a case, primarily, for the goodness or perfection of the Living World, anymore than I'd make a case for the goodness of a Stealth Bomber, if I was trying to explain the *intelligence* and *purposefulness* that it embodies. Remember, a Stealth Bomber, an extremely clever piece of technology, is a *far less* intelligently complex mechanism than *any* single Living Cell, even including those tiny Living Cells going all the way back to the very beginning, 3.8–3.5 billion years ago – so-called 'simple' microbes of breathtaking complexity that make the clever complexities of any Stealth Bomber look like a baby's rattle in comparison.

'My argument, Ali, is that Nature is amazingly *intelligent, not* purposeless and random.'

'In other words you're on the idealist or 'super' Naturalist side of the argument?'

'Yes, especially as all the materialist, *no-intelligence-involved, 'Life-is-just-accidental-chemistry'* understandings of evolution, including the Darwinian one, try to do the *impossible* – like pulling yourself up by your shoelaces. They look at, say, tiny Living Cells, here from the beginning, amazingly complex mechanisms in their own right, and assume that they are the incredibly cleverly complex results of entirely unintelligent and chance processes.

'This not only doesn't make sense it just doesn't work. And the fact that, currently, a dogmatic, 'pre-Galilean-the-Sun-orbits-the-Earth' style consensus in biology believes that it does work does not make it true. It just makes conventional science and biology look unreasonable. It is, apart from anything else, *in total conflict* with everything we know about causation. Extraordinary claims, as Carl Sagan said, require extraordinary evidence. Yet there is no extraordinary evidence that 'chemical accidents' are what Life is.

'People have spent decades *intelligently* designing experiments to create Life *'by accident'* and all they have ever achieved is to produce some of Life's basic building blocks,[19] that is amino acids. This is like 'accidentally' producing a few bricks, and then arguing that: 'Look, this proves that Rome (or the Living World) arose by accident!' Materialism has made our culture irrational. 'Accidents' don't build clever mechanisms of any kinds, let alone machines. That's not saying anything clever, Ali. It's humanity's *universal experience.*

'We'd not dream of ascribing a Supercomputer to random evolutionary processes but when it comes to Living Nature, all of whose mechanisms are mind-blowingly clever, we seem, as a western culture, to have taken leave of our intellectual and intuitive senses and to believe, with no evidence at all but anti-empirical assumptions, counterfactual presuppositions and dogmatic assertions that lots of mindlessness + lots of unintelligence + lots of time + lots of entropy + lots of accidental chemistry could ever possibly be the

---

[19] Wells, J. 2002. *Icons of Evolution.* Regnery; pp. 81–111. Meyer, S. 2009. *Signature in the Cell: DNA and the Evidence for Intelligent Design.* Harper One.

blind and purposeless causes of the *blazingly clever,* 24/7–dynamic, digitally CODED, Life–Force driven bio-functionalities of our amazing Living World. Science was supposed to be an empirical and evidence based activity.'

'So why do scientists think this way?'

'Mainly because materialist thought sees Nature as unintelligent and accidental. Ideological–materialism opts, as it were, for a 'mindless Mystery' of existence whereas philosophical idealism, currently the minority view in western science, opts for an 'intelligent Mystery' of existence. Idealism reasons that we are conscious, we have *minds and intelligence* and, therefore, so must the multi-dimensional, multi-vibrational Cosmos from which we emerge. As Antony Peake, a contemporary scientific–idealist thinker puts it:

'Self-awareness within *a seemingly unaware universe simply does not make sense,* and psychic phenomena similarly present anomalies that may be the pointers to a hitherto unknown model of science; *a model in which mind creates matter rather than matter creates mind.*'[20]

'That sounds very interesting, Ollie, but *how* can the universe be 'aware'?'

'No one can answer that question. No one actually knows *how* we come to be 'aware' or how anything comes to be 'aware'. The materialist view is that only people and animals have any kind of awareness and that this is as a result, of entirely mindless and accidental processes. The materialist position is equivalent to some desert islanders examining an unknown TV (the body-and-brain in the anology) who assume that the TV is an 'accident' that somehow 'accidentally' mined, refined, manufactured and assembled itself and 'accidentally' made the TV Shows (the intelligent awareness, mind and Soul in the metaphor) as opposed to merely transmitting them.

'In reality no one has a clue where the Universal Witnessing Awareness, the Self-hood, Subjectivity and Sentiency, however primitive or advanced, that all humans and animals share, and plants in various ways too,[21] comes from. Like the desert islanders with the unknown TV, our materialistic science and biology assume total Reality

---

[20] Peake, A. *The Out of Body Experience: The History and Science of Astral Travel.* Watkins Publishing.

[21] See, for example: Tompkins, P & C. Bird. 1973. *The Secret Life of Plants.* Harper Perennial.

to be mindless and unintelligent and therefore that our minds and intelligences 'must', logically, be the 'blind chance' products of 'accidental chemistry'. Modern research shows, however, that human consciousness is not, ultimately, dependent on the body,[22] but can exist beyond it, and this points to the classical, idealist, 'mind-before-matter-and-mechanism' understanding of existence that consciousness, intelligence, desire and purpose are all ontological properties of the fabric of reality, they are not random byproducts of matter.

The world's earlier peoples and today's indigenous peoples have always been aware of this reality and have never lost touch with it. For example in his book, *Waking From Sleep,* Steve Taylor says:

> '... to indigenous peoples there are no such things as "inanimate objects". ... they sensed that even rocks, the soil and rivers had a being or consciousness of their own. ... As the anthropologist Tim Ingold writes, to indigenous hunter-gatherer peoples, the environment "is alive". As a result, says Ingold, hunter-gatherers seek to maintain a harmonious relationship with the environment, "treating the country, the animals and plants that dwell in it with due consideration and respect, doing all one can to minimize damage and disturbance." Or as the Cherokee Native American scholar, Rebecca Adamson points out, for indigenous peoples "the environment is perceived as a sensate, conscious entity suffused with spiritual powers through which the human understanding is only realized in perfect humility before the sacred whole"...
>
> '... indigenous peoples' sense of the "aliveness" of the world goes together with a sense that natural phenomena have their own *being* too, a subjectivity beneath their surface reality. This is what the aboriginal Australians referred to as the "Dreaming" of things. As anthropologist Robert Lawler notes, "Every distinguishable energy, form or substance has both an *objective* and a *subjective* expression." ['tails' and 'heads', chemicals + intelligence, Life + Form, Soul + mechanism]. Aborigines have complained that the problem with European Australians is that they can only perceive the surface reality of the world and can't enter its interior life. As one aboriginal elder stated, "Unless white man learns to enter the

---

[22] See, for example: Kelly, E. F. & E. W. Kelly, A. Crabtree, A. Gauld 2009 *Irreducible Mind: Toward a Psychology for the 21st Century.* Rowman and Littlefield Fontana, D. 2005. *Is There an Afterlife?: A Comprehensive Overview of the Evidence.* Schwartz, G.E. Phd. & W. L. Simon. 2002. *The Afterlife Experiments: Breakthrough Scientific Evidence of Life After Death.* Lommel, P. 2010. *Consciousness Beyond Life, The Science of the Near Death Experience* and Carter, C. 2010. *Science and the Near Death Experience.*

dreaming of the countryside, the plants, and animals before he uses or eats them, he will become sick and insane and destroy himself.[23]

'But isn't the skeptical, western, materialist reply simply that those peoples are all imagining things because they're afraid of thunder and lightning and they haven't mastered Nature the way we have.'

'You mean they still see the Soul and intelligence in Nature. They still say 'please and thank you' to the Nature that births them and sustains them, rather than just grab, take, make and break. You mean that they haven't learn't to dominate and to destroy, to bite the hand of the Nature that gave rise to them saying it's unintelligent, Soulless and meaningless. They haven't yet learn't to poison the land with pesticides, to deplete the soil, cut down the forests, to pollute the air, rivers and seas and to imprison the cattle and poultry in vast sheds causing immense suffering in the animal kingdom? Our culture has mocked such ways of knowing and connecting with Reality because the only consciousness that our science recognizes is a kind of cold, separative, reductive, choppy, mechanical way of thinking. Our science is clever but it often contains little wisdom and little heart.

'It is not just our minds that need to change, Ali, but also our hearts and feelings if we are ever to be able to reconnect with the inner spirit and meaning of things and to realize, not just intellectually but also emotionally and spiritually, in felt, experiential ways, that the Universe really is alive and singing, that that's not a metaphor, it's the truth. It is, however, only a truth that we can now know and recapture through inner shifts, changes in mind, heart and feeling and through inner training and development. Our cold, sometimes even heartless, 'head-based-alone' ways of thinking and apprehending reality won't get us there. We can keep on working with radio telescopes and super colliders, looking into outer space and quantum space, but that also won't get us there. Only by changes of mind, and by 'going within' and transforming from inside out can we make these shifts in experience and understanding.

'The materialist thinking that controls western science and biology, today, obviously dismisses all this. Far from perceiving the Cosmos as conscious and intelligent, it sees it as permeated at all levels and dimensions by 24/7 mindlessness, insentience and unintelligence.'

'What about our consciousnesses and intelligences, though?'

---

[23] Taylor, S. 2010. *Waking From Sleep*. Hay House, p. 44. Text in square brackets added.

'Materialist thinking believes that those are freak 'accidents' that permit us just enough consciousness to be able to do intelligent science and to discover how allegedly unintelligent and 'accidental' are all the workings and functionalities of Nature. Hence atoms, molecules, gravity, light, electricity, Life, DNA CODING are all 'accidents', meaningless and purposeless functionalities – regardless of the oxymoronic nature of the position. Because of this one of the world's most famous and important materialist philosophers, Daniel Dennett, has even held that:

"We're all zombies. Nobody is conscious."[24]

'So who does he think is speaking when he says that?'

'Ali, I'm not entirely sure. However, modern science and biology far from demonstrating, as materialist thought seems to wish, that Nature is 'unintelligent', have shown – as Aristotle, Bacon, Galileo, Newton, Faraday and all the other founding giants of science would have predicted – quite the opposite. That actually the inner workings of Living Nature are fantastically *intelligent* regardless of the precise nature of its mysterious sources, our reservations about the world's religions, our bafflement at the challenges of existence for all Living Beings or the meaning of the existences of infectious diseases, parasites, wasps, poisonous spiders, snakes and so on.

'We are not making a moral judgment about these amazingly clever expressions of global Nature's intelligence, which perhaps contains 'dark' as well as 'light' forces. After all, the cleverness of Living Nature's mechanisms is one thing but as to the intelligences behind the myriad existences it gives rise to they are another.'

'How do you mean?'

'I mean that we live within a multi-dimensional, multi-vibrational Cosmos of intelligence and meaning, not one of mono-dimensional mindlessness and accident alone as materialism believes. But intelligence and meaning do not always imply 'all sweet, rosy, good and kind'. We don't yet know precisely what subtle, inter-dimensional laws and forces are involved in the intelligent design or evolution, through vast time, of poisonous spiders and snakes, for example, but we won't solve that mystery by claiming that such amazing and magical Soul Life Form Beings are the purposeless and meaningless effects of unintelligent and 'accidental' chemistry anymore than we'd pretend that a Stealth Bomber, vastly less cleverly complex than any snake or spider,

---

[24] Dennett D. 1992. *Consciousness Explained*. Back Bay Books, p. 406. Quoted by L. Dossey: 2010. Is the Universe Merely A Statistical Accident? *The Blog. Huffington Post*.

was the meaningless and unintended effect of various unintelligent and chance based processes.

'A Stealth Bomber's moral merits are quite apart from the focused *intelligence* and *purpose* that is needed for it to arise and take form out of rocks and stones, to be idea-lized, organ-ized and material-ized by incredibly *intelligent* forces – humans – who, as amazing spiritual Beings themselves, temporarily physical-ized and local-ized in space and time, also come out of rocks and stones – by *astonishingly clever processes,* not 'accidents'; processes driven by 24/7–dynamic laws and forces that are vastly more sophisticated than the ones that we, clever as we are, use to get Stealth Bombers out of the rocks and stones and into the sky.

'In *Billions of Missing Links: A Rational Look at the Mysteries Evolution Can't Explain,* G.S. Simmons lists dozens of processes and organisms that cannot be explained or explained away as randomly evolved, Darwinian–style accidents. Here is just one, and it's also a good example of Michael Behe's famous principle of *irreducible* complexity. As he explained in his world-class book, *Darwin's Black Box,* irreducible complexity describes the fact that countless mechanisms in Nature including the irreducibly complex bacterial flagellum, the irreducibly complex blood clotting cascade and many other aspects of intra cellular machinery (just as most or all of the mechanisms we make), cannot arise gradualistically in some kind of incremental, random micro step by random micro step Darwinian way of slowly rising complexity because *all their parts* have to arrive and be put together at the same time or they'll simply not work at all – as with even the most simple mouse trap, they are irreducibly complex.

'Simmons says, of the irreducibly complex Bombardier Beetle:

'This African insect can fire off two chemicals, hydrogen peroxide and hydro quinone, from *separate storage tanks* and rear jets. When the chemicals combine, they form a new chemical that burns the predator. The beetle can shoot these chemicals with an uncanny accuracy, as well, to either side, backward, or even forward, by swinging its tail under its abdomen. Special nozzles blast predators at a rate of 500 bursts per second, each at a speed of 65 ft./s. These chemicals are potent enough to severely damage a mouse and injure the eyes of any animal … Yet these chemicals are entirely benign when stored *separately* at the back end of these beetles. How could this happen by accident? "Oops, those two chemicals didn't work" (spoken by an intermediate species). "Mind if I try two others before you eat me?" Or, "Could you stand a little taller so I can get you with my nozzles?"

Keep in mind there are hundreds of thousands of chemicals on this planet to choose from. And even if the combo turned out perfectly right the first time, the beetles still needed a way to make them, store them, and fire them off.[25]

'So examples like these refute the Darwinian 'Life-is-a-randomly-evolved-accident' theory?'

'Yes, of course they do. They are *irreducibly complex all or nothing processes* and *there is no extraordinary evidence anywhere to give any support at all* to the extraordinary materialist claim that creatures such as the Bombardier Beetle are the random, freak products of unintelligent, blind-chance-based evolutionary processes. However, if you present Darwinian thinking with such examples, the attitude will be to mock and to label you a 'creationist' as though you're simple minded for pointing out that these processes cannot be explained or explained away as 'accidents' of unguided, Life-Forceless chemistry, no matter how much some people seem to like that idea and its associated one of a Cosmos of utter mindlessness and unmeaning.'

We walked on through the beautiful meadows and fields, beside our companion, the refreshing river.

---

[25] Simmons, G. 2007. *Billions of Missing Links.* Harvest House, p. 29.

3 Materialism: Life is an Accident – But is This Really True?

> '[The] break between the living and nonliving world... [is] the *most dramatic* and fundamental of all the discontinuities of nature. Between a *living cell* and the lowest highly ordered *non-biological* system, such as a crystal or snowflake, there is *a chasm* as vast and absolute as it is possible to conceive.'[26]

After continuing on quietly for a while, I said, 'Humanity, Ali, is actually very well equipped to use the Soul quality we call *intelligence* to work out the difference between things that have *intelligent* causes and ones that don't.

This is, in essence, all that the modern theorists of intelligent design or ID in America and elsewhere have claimed to a great deal of mockery from materialist and Darwinian dominated science and biology. It's not as though the ID theorists are saying we should stop doing science just because Nature, it turns out, is intelligent and not 'idiotic' after all but simply that certain features of the Living World point to *purposeful* not 'accidental' causation. Accepting this possibility doesn't mean we are a step away from the end of science.

'Most of the world believes in some kind of higher power or intelligent Higher Nature so it begins to look like materialist special pleading to protest that science is incapable of determining whether what the majority of humanity already *rationally infers,* logically deduces and intuits to be true, just as all the founding giants of science did, is true or not.

'What do you mean?'

'I mean that scientists can put forward the hypothesis that Living Nature is driven by intelligent and purposeful processes, as opposed to purely mindless and accidental ones, regardless of their physical invisibility and their precise nature or motives – which is the materialist hypothesis of Nature, Life and Mind.

'Look, no one, today, imagines that the cows and sheep in these fields or the birds flitting to and fro were made out of spit and clay by some anthropomorphic bearded-god anymore than anyone thinks that this randomly evolved river valley was scooped out by a giant's hand.

---

[26] Denton, M. 1985. *Evolution: A Theory in Crisis.* Burnett Books, p. 249. Emphasis added.

These realizations don't, however, give us a good reason to assume that the same kinds of accidental processes, involving just earth, rocks and water that *unintelligently* evolve this river valley could randomly evolve these cows and sheep and birds, whose amazing, 24/7 bio-functionalities make any clever stealth bombers, supercomputers or space shuttles we make look like wooden spoons in comparison.

'Even long after Darwin published there were always scientists and thinkers who weren't persuaded by his theories. Eventually, even his world famous co-inventor of the theory of Natural Selection, Alfred Russell Wallace, turned against the Darwin–Wallace theory, as it was initially called, in its most extreme, that is in its most materialistic forms.'

'Why did he turn against it?'

'Because with the passage of time Wallace concluded that his and Darwin's theory of random, Soulless and Life Forceless blind-watchmaker evolution did not fit the biological data. He realized, for example, that *none* of his or Darwin's ideas on natural selection could rationally account for the amazing Life Force driven transformation that a caterpillar undergoes in its metamorphosis into a butterfly. He saw that there were no *rational* Darwinian, naturally selected 'accidental' stages that could account for this astonishing and *irreducibly* complex, Life and Soul Force driven, and all or nothing process – Egg – Caterpillar – Cocoon – Chemical Soup – Butterfly! This radical insight applies to *all* the Life-Force driven transformations we see in Nature. Larval form to adult insect. Egg to Tadpole to Newt or Frog.

'Today, in conflict with Darwin and Dawkins, many other scientists have come to the same conclusion as Wallace. Not only this, but they have now begun to show, *empirically and scientifically,* that even if we passionately *want* the Living World to be the amazingly 24/7–clever result of random chemical processes, as many materialist and Darwinian thinkers seem to do, *it just doesn't work*. Living Nature *could not possibly* have arisen in that way as the astonishingly smart result of myriads of allegedly unintelligent and 'accidental' processes. The materialist view of Living Nature *never* really made sense but its advocates always believed that while it was highly improbable it was just about possible. Today, people have begun to realize it's impossible. This is what Sir Frederick Hoyle, the famous astronomer, said about this:

> 'Life **cannot** have had a random beginning ... The trouble is that there are about two thousand enzymes, and the chance of obtaining them all in a random trial is only one part in $10^{40,000}$, an outrageously small probability that **could not be** faced even if the **whole universe** consisted of organic soup.'

'He added,

> 'Once we see, however, that the probability of life originating at random is so utterly minuscule as to make it absurd... It is therefore almost inevitable that our own measure of intelligence must reflect ... *higher intelligences* ... even to the limit of God ... such a theory is so obvious that one wonders why it is not widely accepted as being *self-evident*. The reasons are *psychological* rather than scientific.'[27]

'The existence of *any* kind of Cosmos is a mystery, for why should anything at all exist rather than nothing at all? The question is what does exist and *what kind* of Cosmic mystery are we in? An essentially intelligent one, in which many accidents happen without, however, random events being the fundamental drivers of existence, or a totally mindless and accidental one? Saying that the existence of matter is not a 'mystery' because it appeared as a result of a Big Bang doesn't answer that question. It merely regresses the mystery.

'If a never-seen-before Rolls-Royce appeared on a desert island some of the islanders could argue – as materialist science does today, in respect of the Big Bang, the Universe and our Living World – that it's an *unintelligent* mystery, a random event of some kind. Of course, they'd be wrong. The clever Car's mystery (as a metaphor for the Big Bang, the Universe, the Living World) would remain to be explored, as our sciences do, but it would be an *intelligent* mystery, not an unintelligent one. The island's scientists would be the dynamic, curious, inquiring *intelligences* exploring a manifestation of *intelligence* – the shiny Car (or Living World). Materialist thought, however, rules out any possibility that Nature's ultimate sources may be intelligent and that scientists are perfectly capable of testing that hypothesis.

We walked on quietly for a while, then Alisha spoke.

'One thing that intrigues me, Ollie, is why was Darwin *so certain* that the Living World was not a manifestation of intelligence and purpose but rather some kind of meaningless and unintended chemical 'accident'?'

'That's a good question. Darwin believed that Life 'had to be' just some kind of long running *'accident'* because Living Nature was, it seemed to him, so wasteful, cruel and harsh. He couldn't see how the simplistic, 'good-father' god of the Victorian imagination could be responsible for what *he* saw as Living Nature's evils. In a letter to a friend he wrote:

---

[27] Hoyle, F. and N.C. 1981. Wickramasinghe, *Evolution from Space*. J. M. Dent & Sons.

"What a book a devil's chaplain might write on the clumsy, wasteful, blundering, low, and horribly cruel works of nature".[28]

Yet, 'cruelty, waste and mess' are all psychic and moral judgements that Darwin's caring, thinking mind and Soul permitted him to make. Do the supposedly unguided atoms of oxygen, carbon, hydrogen and nitrogen, out of which our allegedly 'purposeless', *'accidental-chemical-reaction'* bodies are formed make such moral judgements? Do the atoms, amazingly smart-materials as they are, care about such things?

'No, of course not, Ollie. It seems clear to me that Mind and Soul, Sympathy and Antipathy, Intelligence and Purpose are things apart from matter even if they can, at the same time, pervade it and in-form it. After all TV Shows are apart from TVs yet for us to see them they have to pervade the TV and to be transmitted by the TV.'

'I agree, but materialist thought which is sometimes referred to as methodological naturalism (where total Nature is conceived to be entirely material and without any intelligently informing and organising Life and Soul Forces), hypothesizes that the Living World is like a random TV that 'accidentally' mined and refined itself out of the ground and 'accidentally' began not merely *transmitting* TV Signals and Shows (Minds and Souls), but 'accidentally' *making* the TV Shows as well.

'In 1871, pursuant to this kind of purely materialistic hypothesis of Life and Mind, as mere 'accidents of chemistry,' Darwin wrote to his friend, Joseph Hooker, "But if (and oh what a big if) we could conceive in some warm little pond with all sorts of ammonia and phosphoric salts – light, heat, electricity etc present, that a protein compound was chemically formed, ready to undergo still more complex changes" – and, Darwin implied, become a tiny Living Cell, *"the most complex system known to man."*[29]

'Darwin thought, based on the then available scientific knowledge, that Living Cells, smaller than sand grains, were *relatively simple forms of chemical glob,* and, therefore, that they could have arisen by fairly simple and random processes. Haeckel, one of Darwin's contemporaries, described the Cell as a "homogeneous globule of protoplasm". The Victorian materialists could not have been more wrong. Living Cells are literally, scientists now realize, like fully automated, 24/7, micro-miniaturized chemical factories. As Michael Denton says:

---

[28] Hunter, C.G. 2002. *Darwin's God: Evolution and the Problem of Evil*. Brazos Press.
[29] Davies, P. 1998. *The Origin of Life*. Penguin.

'Although the tiniest bacterial cells are incredibly small... each is in effect a veritable micro-miniaturized factory ... *far more complicated than any machinery built by man* and absolutely without parallel in the non-living world.'[30]

'Now this is also really important, Ali:

'Instead of revealing *a multitude of transitional forms* through which the evolution of the cell might have occurred, molecular biology has served only to emphasize the **enormity of the gap** [between non-living and living matter].

'He continues:

'We now know not only of the existence of a break between the living and non-living world, but also that it represents the *most dramatic* and fundamental of all the discontinuities of nature. *Between a living cell* and the lowest highly ordered *nonbiological* system, such as a crystal or snowflake, there is *a chasm* as vast and *absolute* as it is possible to conceive.'[31]

'In other words, Oliver, there's no evidence that Living Cells, at the base of all the larger Life Forms, gradually evolved out of earth, rocks and water, perhaps by way of crystals, in a blind and random yet step by step, Darwinian way of gradually rising complexity?'

'Yes, exactly. He's also saying that the difference between sentient, Living Beings and the rocks and water in the ground is vastly greater than the difference between the rocks and stones in this river and a Space Shuttle or Supercomputer.'

'But they are all made of the same materials aren't they, Oliver? All Living Cells and our Bodies come out of these rocks and stones and water and so do Space Shuttles and Supercomputers.'

'Yes, of course. But Space Shuttles and Supercomputers don't come out of the rocks and stones 'by accident', Ali. It takes amazing human *intelligence* to produce such mechanisms. Yet these incredibly clever devices are utterly simple compared to the supercomputer–speeded, living complexities of even the very earliest and so very misnamed 'simple' single-cellular organisms, bacterial cells here, all perfect and complete, from the very beginning, 3.8 billion years ago.

---

[30] Denton, M. 1985. *Evolution a Theory in Crisis*, Burnett Books, p. 250. Emphasis added.
[31] Ibid., pp. 249–250. Emphasis added.

'The point is, classical and modern intelligent design thinking both argue that *intelligent* laws, forces and processes, not mindless and unintelligent ones, are behind and give rise, right now, second by second, to Living Nature. Materialist or unintelligent design scientific thinking, on the other hand, argues that there simply *are* no intelligent forces within or behind Living Nature and that what we and everyone else in the world *are* and what we are *intelligently and mindfully* observing, right now, this very second and every second, for example these sweet meadow grasses, dandelions, buttercups, sheep and cows is a form of Life-Forceless, unintelligent and *accidental chemistry* – that we are all, quite literally, *'accidental chemical reactions!'* – walking zombies basically. As Dennett one of the world's most famous materialist philosophers said:

"We're all zombies. Nobody is conscious."[32]

'That seems irrational, though, Ollie. Does anyone really think that?'
'I agree, it's a peculiar view of Living Nature, but, it is the mainstream scientific and biological view today – very Life itself is supposedly an ongoing, *'accidental chemical reaction'*. Methodological–materialism or methodological-naturalism has captured classical western science and made parts of it irrational and false to reality.
'What do you mean by *methodological* materialism?'
'Methodological materialism is the widespread and governing assumption in science and biology today that, when doing science, scientists must assume that material causes alone – (that is just incredibly smart-materials-chemicals (which in materialist thought are assumed to be accidents) + stunningly clever chemical Laws (which are also supposedly accidents) + 'accidental', unintelligent, Life-Forceless and Soulless combinations of these) – are fully sufficient to explain all Natural phenomena including the Living World and the Minds and Intelligences of the scientists intelligently pursuing science to explain how unintelligent, purposeless and 'accidental' Nature is. Methodological materialism is also sometimes confusingly, and misleadingly, referred to as methodological naturalism.
'Why is that misleading?'
'It's misleading, and confusing, because it assumes, before it even starts, that Nature is entirely material, just chemicals, and that it contains no intelligent or 'super' or 'higher' elements. Yet it is the very nature of Nature, the true character of Nature that is the key question.

---

[32] Dennett D. 1992. *Consciousness Explained.* Back Bay Books, p. 406.

To dogmatically assume, as a matter of science, that Nature is material alone excludes from the start the scientific possibility that there may be more to Nature, especially to Living, Sentient, Conscious Nature, than just physics and chemistry. That's a real knowledge blocker and science stopper.'

'Why?'

'Because, by definition, it excludes the possibility that modern scientists are capable of investigating precisely that hypothesis and of determining whether or not it may be true.'

'You mean . . .'

'The scientific hypothesis that there may be more to Nature, especially to Living Nature and Consciousness and Mind, than the known laws of physics and chemistry and that it may be possible, today, to demonstrate this empirically – as a matter of genuine science and not mere philosophy or superstition. It is precisely this hypothesis that the modern theorists of intelligent design such as Behe, Dembski and Meyer seek to investigate.

'Yet you say that the standard scientific view is that we have to assume, when doing science and biology, that just allegedly accidental chemicals + equally accidental chemical laws + a large number of chemical accidents are fully sufficient to explain the amazingly dynamic functioning that is Living Nature?'

'Yes, that is what methodologically materialist and methodologically naturalist thinking both are.

'But the methodologically materialist viewpoint assumes, as a matter of science, the very thing that it has not actually demonstrated – that total Nature, or all that is, is matter alone and, surely, Ollie, it is, in any case, clearly self-refuting. Living Nature's 24/7–dynamic mechanisms are obviously and astonishingly smart. It's a distortion of language and meaning to claim otherwise.'

'Yes, that's the point. This is why methodological materialism is a millstone around modern biology's neck because it closes off scientific inquiry. Materialism and its non-negotiable spirit of anti 'super' Naturalism, of anti-intelligence-or-purpose-in-Nature, as a matter of 'science' is a science blocker.

'The problem is that far too many people have been fooled into thinking that science and the ideology of materialism, which is just a particular world-view, are one and the same. Genuine science, however, simply describes a vast range of methods for investigating the true nature of reality, of all that is, be it physical or non physical, obvious or subtle, material or spiritual. The fact that it is, in many ways, easier to investigate and to manipulate physical reality (what we tend to think of as take-it-for-granted-it's-nothing-so-very-special 'ordinary' Nature as opposed to subtle, intelligent 'super' or 'higher' Nature), does not mean that other more subtle and non-physical elements and dimensions of total Nature, of all that is, cannot also be studied scientifically – that is in systematic and disciplined ways as, in fact, a minority of several generations of courageous and open-minded scientists have been doing for over a century now.

'What I don't understand, though, Ollie is how did methodological materialism or so called methodological naturalism ever take off because they both seem so obviously counterfactual to the true nature of Nature? Materialism – or a question begging 'naturalism' that references chemical matter alone – surely, cannot account for Life, let alone for Mind and Consciousness, Sentiency and Subjectivity and all the Soul qualities, Love, Purpose, Intelligence, Laughter and Fear, Sadness and Joy ...'

'Look, it took off in the west, because, starting around 200 years ago, more and more people rightly realized that the world couldn't possibly have been made by some kind of simplistic, anthropomorphic, bearded-god figure, out of spit and clay, in 6 busy days 6,000 years ago and be responsible for a vast universe, one which, we now realize, is 20 billion light years across. And people also thought, with far less reason, that if there is no such entity then our Living World must be the incredibly intelligent (seeming only) product of mindlessness and trillions of 'accidents'.

'The trouble is that a very simplistic narrative of 'super' Natural or *intelligent Cosmic* causation was replaced by an even more naive materialist narrative of *unintelligent and accidental* causation.'

'Why was it even more naive?'

'Because the 'bearded–god', six day creation story was only ever a metaphor. It did, however, understand that intelligent mechanisms only ever have intelligent causes and that like produces like. Random floods and rain produce randomly evolved river valleys. Copious inputs of intelligent information produce intelligently evolved cars, TVs and computers. While it's obvious to us, given the ancient age of the Earth and the pattern of Life's unfoldment over deep time, that Living Nature could not possibly have been made out of spit and clay in six busy days, a few thousand years ago, it should also be obvious that it cannot possibly be a billions year long running *'accidental chemical reaction'* that expresses, 24/7, by stunningly clever processes, as amazing Bacteria, Turnips and Hares, Parsnips and Bears, Lions and Dolphins, Eagles and People – there is nothing in the laws of Life-Forceless, zombie, 'accidental' chemistry to warrant such a conclusion.'

We walked on quietly. Eventually, I said,

'Michael Denton says that "even the simplest of all Living systems on earth today, bacterial cells" are *stunningly* complex Life Forms, almost complex beyond belief. He explains that:

> 'Molecular biology has also shown that the basic design of the cell system is essentially the same in *all* living systems, from bacteria to mammals. In *all* organisms the roles of DNA, mRNA and protein are identical. The meaning of the genetic code is also virtually identical in all cells. The size, structure and component design of the protein synthetic machinery is practically the same in all cells. In terms of their basic biochemical design, therefore *no living system* can be thought of as being *primitive or ancestral* with respect to any other system, *nor is there the slightest empirical hint* of an *evolutionary sequence* among all the incredibly diverse cells on earth. For those who hoped that molecular biology might bridge the gap between chemistry and biochemistry, the revelation was profoundly disappointing.[33]

'In other words, he's saying that *all the Cells that have ever lived*, out of which *all* the larger Life Forms are built, be they single cells of ancient non-nucleated, prokaryotic bacteria, trilobites built out of nucleated, eukaryotic cells, ammonites or dinosaurs or modern fish, frogs, birds, butter cups, beech trees, apple trees or people are all *very similar* in terms of their incredible levels of complexity and the way they function – there is **no evidence** for any kind of Darwinian-style *progressive sequence of gradually mounting complexity in their arising*.'

'So, at least as far as Living Cells are concerned, the Darwinian idea of gradually rising complexity driven by random processes and chemical accidents does not stand?'

'Yes, that's the point. *It stands refuted.* The earliest, so called 'simple' Cells are extraordinarily complex mechanisms, Living, Self-Reproducing micro-miniaturized chemical factories. In Darwin's day people thought Cells were fairly simple globules of protoplasm. Today, it's unreasonable, or even deluded, to continue with the purely materialistic or unintelligent design hypothesis that this kind of amazingly complex and dynamic bio-functioning is just some kind of Life–Forceless, 'it's-not-smart-at-all', 'chance chemistry'. There is *no genuinely good, evidence-based reason* that anyone has ever found to believe that Living Cells are the incredibly clever results of unintelligent, random processes, which is the view that has dominated science and biology since Darwin.

---

[33] Ibid. p. 250. Emphasis added.

'So why does any one believe it, Ollie?'

'Because of the materialist line that goes: "We can't see any super or higher Natural laws, forces or processes or agencies anywhere, not even in the best telescopes or microscopes, can you?"'

'But such forces can be *rationally inferred,* or intuited, can't they?.'

'I know, but this is what many people came to think and still think.'

'Why though?'

'I think partly because the organized religions' narratives of existence seemed to provide so little by way of understanding of the enigmas of existence, such as suffering and disease, what really happens when we die, what's really the point of it all and so on. This doesn't mean that these religious traditions were of no value but that they no longer answered to modern people's needs which were for *more and deeper knowledge* about reality, both physical and multi–dimensional, Holistic Natural reality. What is Life after death actually like, for example? There has been a lot of research, over the last one hundred and fifty years, which has gradually built up a great deal of knowledge in this area, to the extent that it is now quite clear (to those prepared to make the effort to review just a small fraction of the voluminous literature on the subject) that 'death' is not the end, rather it is an inter–dimensional transition.[34]

'However, most of the western religions instead of embracing and showing some interest in the extensive, modern, psychical and spiritual–science research, providing, by now, very good evidence of the survival of individual consciousness and of Life after death[35] showed very little interest in this modern knowledge and research but just kept to their old dogmas about 'heaven' and 'hell' etc. This contributed to turning people off traditional western religion, (principally Judaism, Christianity and Islam) and drove them into the waiting arms of materialist scientific thought which, at least, made a good job of explaining how the materials and mechanisms of outer Nature worked even if it hadn't a clue as to its inner or subjective workings (its Aristotelian, formal and final causes) and contented itself by simply claiming that there were none and that minds and intelligence were 'accidents' of atoms. Yet, in the context of our knowledge of the capacities of 'accidental' causation (randomly evolved river valleys) and the capacities of intelligent causation (intelligently evolved technologies),

---

[34] Fontana, D. 2005. *Is There an Afterlife?: A Comprehensive Overview of the Evidence.* O Books. Carter, C.2012. *Science and the Afterlife Experience.* Inner Traditions.

[35] Schwartz, G.E. Phd. & W.L. Simon. 2002. *The Afterlife Experiments: Breakthrough Scientific Evidence of Life After Death.* Carter, C. 2010. *Science and the Near Death Experience.*

the materialist explanation of Living Nature doesn't make sense. Molecular biologist Michael Denton says:

> 'To grasp the reality of life as it has been revealed by molecular biology, we must magnify a cell 1000 million times until it is 20 km in diameter and represents a giant airship large enough to cover a great city like London or New York … On the surface of the cell we would see millions of openings, like the portholes of a vast spaceship, opening and closing to allow a continual stream of materials to flow in and out. If we were to enter one of these openings we would find ourselves in a world of *supreme technology* and bewildering complexity. We would see endless highly organized corridors and conduits branching in every direction away from the perimeter of the cell, some leading to the central memory bank in the nucleus and others to assembly plants and processing units. The nucleus itself would be a vast spherical chamber more than a kilometer in diameter, resembling a geodesic dome inside of which we would see, all neatly stacked together in ordered arrays, the miles of coiled chains of the DNA molecules. A huge range of products and raw materials would shuttle along all the manifold conduits in a highly ordered fashion to and from all the various assembly plants in the outer regions of the cell.'[36]

'Every Living Cell, of which we have trillions in our bodies, a little microscopic Life in its own right, a "world in a grain of sand", is *like a micro-miniaturized factory* on the scale of a city. Cells which make the complexity of many of the larger organs, like kidneys, stomach, intestines, ears, nose and so on, that they help to build, look simple in comparison – and they make *all* the things we make and the ploddingly slow time it takes us to make them, look utterly simple minded. Yet, there is no ordinary let alone any extraordinary evidence that such dynamic Living complexity is the stunningly clever product of unintelligent and 'accidental' forces.

'Are you saying that the individual cells that build the stomach or the kidney are far more complex than the stomach or kidney itself?'

'Yes. Those large organs are, as systems, much more simple than every one of the millions of little Cells that go to build them up. This is the point, Living Cells, tiny as they are, are the most complex components of the Life Forms that they help to build, and they have been here from the beginning, billions of years ago, with *no signs of gradual, random, Darwinian–style rise to ever increasing complexity*, which is the essence of Darwinian thinking.

---

[36] Denton, M. 1985. *Evolution a Theory in Crisis*, Burnett Books; p. 328. Emphasis added.

'Michael Denton says that for humanity to build a scale model of *just one* nucleated, 'spaceship' Living Cell, magnified a thousand million times, around 20 km across, larger than inner London or New York, out of 10,000,000,000,000 tennis ball sized 'atoms' at one atom per second would take us *thousands of years* to build. Yet, our spaceship–factory Cells do this in just a few hours.'

'That does sound amazing . . .'

'That's the point, it is amazing and it is intellectually unreal to continue with the methodologically materialist hypothesis that such processes are 'nothing special', there's no intelligence or purpose or teleology to them, they're just 'accidents'. That's not a rational view of what blazingly clever, 24/7–dynamic Life is.

'The trouble is, after Darwin, rationalism and rational thinking, in a peculiar inversion of classical, western scientific thought, predominantly an idealist tradition until then, became equated with the idea that Living Nature was unintelligent and wholly accidental. And, strangely, those, like Alfred Russell Wallace, who argued otherwise became the 'irrational' ones. Yet, Denton says that if we could enter a Cell:

> 'We would see all around us, in every direction we looked, all sorts of robot-like machines. We would notice that the simplest of the functional components of the cell, the protein molecules, were astonishingly, complex pieces of molecular machinery ... We would wonder even more as we watched the strangely purposeful activities of these weird molecular machines. . . .
>
> 'We would see that nearly every feature of our own advanced machines had its analogue in the cell: **artificial languages** and the **decoding systems**, memory banks for information storage and retrieval, elegant control systems regulating the **auto-mated assembly** of parts and components, error failsafe and proofreading devices utilized for quality control, assembly processes involving the principle of prefabrication and modular construction. . . .
>
> 'What we would be witnessing would be an object resembling an *immense automated factory, a factory larger than a city* and carrying out almost as many unique functions as all the manufacturing activities of man on earth. However, it would be a factory which would have one capacity not equalled in any of our own most advanced machines, for *it would be capable of replicating its entire structure within a matter of a few hours.*'[37]

---

[37] Ibid. p. 329. Emphasis added.

'That is incredible, Ollie, but don't some people say, as a way of explaining cellular complexity, that microbial cells arrived on Earth by meteorite?'

'Yes, they do, but that just regresses the mystery of Life, it doesn't answer it. Philosophical materialism, which controls scientific and biological thinking today, assumes that total reality is 'mindless', 'unintelligent' and 'accidental' and, therefore, if the materialist view is correct, that the Living World must be too. This materialist view of reality, however, for it to be scientific, and not merely a dogma or superstition, *must be open to testing, to falsification.*

'What the advocates and ideologues of the materialist world view refuse to accept is that the materialist hypothesis has been tested and it has been falsified. There are simply too many phenomena that the materialist hypothesis that 'only-mindless-matter-exists' and 'Life-on-Earth-is-an-accident' cannot accommodate. The stunning intelligence, purposefulness and complexity of second by second, biological organization and the modern evidence for extrasensory abilities[38] and for Life after death[39] to name but several. Materialism is false to reality.

'Do you think this is why it's so hostile to the modern, idealist or non-materialist intelligent design movement?'

'Yes, because idealism, non-materialist thinking, takes the view that some features of Nature in general and of the Living World in particular are expressions of incredibly intelligent laws, forces and processes and that it is a distortion of our language to claim otherwise. If we do so we run the risk of making science dishonest and irrational. This is why Denton describes the Cell as he does. He's not saying we should stop doing science because, it turns out, Nature expresses tremendous intelligence but merely that it makes a nonsense of scientific rationality and of our understanding of the Cosmic Mystery in which we find ourselves to keep on assuming, as a matter of science, that the *whole thing* is mindless and unintelligent and, above all, accidental. No one denies that 'accident' plays a part in the dramas of existence. But that is a far cry from the extraordinary and entirely unevidenced materialist claim that Life and Mind and the Living World and all their clever laws, astonishingly smart-materials and 24/7–dynamic processes are accidents.

---

[38] Radin, D. 2009. *The Conscious Universe.* Harper One. Carter, C. 2012. *Science and Psychic Phenomena: The Fall of the House of Skeptics.* Inner Traditions. McLuhan, R. 2010. *Randi's Prize: What Sceptics Say About the Paranormal, Why They Are Wrong, and Why It Matters.* Matador.

[39] Fontana, D. 2005. *Is There an Afterlife?: A Comprehensive Overview of the Evidence.* O Books. Carter, C. 2010. *Science and the Near Death Experience.* Inner Traditions.

4 Codes: Cannot Arise By Accident – It's Impossible

'The basic flaw of all [materialistic] evolutionary views is the origin of the information in living beings. It has never been shown that a CODING system and semantic [meaningful] information could originate by itself... The information theorems predict that this will never be possible. A purely material origin of life is thus [ruled out]'[40]

We walked on for a moment, then Ali said,
'But I'm still baffled by the materialist belief in accidental causation as *the key* 'explanatory' agency of Living Nature, Ollie, when it goes against not only logic but also *all* our real-world experience of 'accidents', what they can actually cause – randomly evolved landscapes or car crashes and spilt milk – and of what *intelligent* causes lead to – TV's, Computers and Life Forms more complex than all such things put together?'

'It is caused, I think, above all by the fact that Living Nature's intelligent, subtle, spiritual or 'super' Natural sources cannot be physically seen but can only be rationally inferred, intuited or, to some extent, directly experienced by means of inner vision or higher–sensory perception such as India's yogi–adept, spiritual–scientists had developed centuries ago.[41] I think it's also partly caused by a strong dislike, in the materialist community, of many of the world's religions and their officials and their tendency to boss everyone around and insist on dogmatic 'truths' for which, in some cases, there's no more evidence than the materialist, unintelligent design claim that Living Cells more complex than Space Shuttles are 'accidents'.'

'But that's not a good enough reason to abandon classical logical thinking is it?'

'I agree. This is why I wanted to explain my ideas to you. The good

---

[40] Gitt, W. 2006. *In the Beginning Was Information*. Master Books, p. 124. Emphasis and text in square brackets added.

[41] For examples of yogic knowledge see, for instance, (1) Davidson, J. 1992. *Natural Creation or Natural Selection?* Element. (2) Tompkins, P. 1997. *The Secret Life of Nature*. Thorsons. (3) Yogananda, P. 1946. *Autobiography of a Yogi*. Self Realization Fellowship (2006). "One of my favorite books. ... I cannot recommend it enough to anyone, any age, any religion, any place, any time!" Amazon reviewer.

news is that many scientists are starting to realize that Life–Forceless and Soulless *'accidental-chemical-reactions'* or ACRs of the kind that arise when lightning strikes pond-water are not what Life and the Living World actually *are*. They have started to see that not only do ACRs *not* provide the all important *Life and Soul Forces* that are the difference between *Living* Soul Beings and *Dead,* Soul-empty, chemical shells but neither can entropic, *chemical accidents* generate the enormous amounts of digital CODING information required to run the Life Forms, they destroy it.

'I notice, Ollie, that you refer a lot to Life-Forceless and Soulless accidental chemistry as being what materialist biology considers *Life* to be? Surely, modern science recognizes a *Life or Vital Force* of some kind?'

'That's just it, it most absolutely does not.'

'Why though? Isn't it impossible to explain Life without inferring the existence of such a Life Force or Vital Force or forces?'

'Yes, I agree. But a methodologically–materialist biology cannot recognize such a force because to do so overturns its paradigm, it *falsifies materialism*.'

'So what's the problem, wouldn't scientists be pleased to realize that Nature is intelligent after all and that Life is something more than physics and chemistry?'

'You'd think so but, sadly today, science and biology are in the devitalizing grip of the materialist dogma that *all phenomena* in Nature, including those of the Living World can be explained purely and mechanistically in terms of Life-Forceless, Soulless accidental chemistry. That, in fact, as Francis Crick put it:

> "The ultimate aim of the modern movement in biology is to explain all biology in terms of physics and chemistry"[42]

'For this school of thought, admitting the existence of the Life Forces and Soul Forces implies 'super' Naturalism or *intelligent* causation of the Living World and it's therefore taboo.'

'But I thought science was supposed to be a logical and reasonable activity, about *hypothesis testing*, and all about what's actually true, Ollie. Surely it's obvious that the Living World is more than second by second, accidental chemistry.'

'It may be obvious to you, Ali, but, sadly, it's not obvious to the materialist thinking that currently controls biology. This dogmatic thought school also insists that chemical 'accidents' are responsible for the

---

[42] Crick F. 2004. *Of Molecules and Men*. Prometheus, p. 10.

creation of the huge amounts of **digitized information** contained in the DNA CODED genomes, that Bill Gates described as "like a computer program but far, far more advanced than any software ever created".[43] This is despite the fact that computer software is (a) never known to arise by accident and (b) modern information theory predicts that CODES **cannot 'arise by accident'** even if we'd passionately like them to.[44]

'The materialist view, however, hypothesizes that the *forces of entropy,* which in classical scientific reality firmly *prevent* the 'accidental' creation of computer-type CODES would not be a problem for Life's evolution because the energy of the Sun would provide the Living World with all the 'free fuel' it needed to drive random evolution and to 'accidentally' develop *digital,* DNA CODED programming software, "far more advanced than any software we've ever created", forward.

'This, though, is as reasonable as imagining that an *accidental chemical reaction* like a random forest fire, with all the 'free fuel' it ever needed could evolve computer languages like COBOL or BASIC rather than anything other than burnt woods. Living Cells, tiny, yet internally vast, micro-miniaturized factories, do *not* depend on 'free' Chemicals + 'free' Energy + 'free' Accidents for their existence but on enormous supplies of intelligent, DNA CODED information. This highly specific, clever, digital information, which runs all Living Cells, is, of course, as W. Dembski points out in *No Free Lunch: Why Specified Comlexity Cannot Be Purchased Without Intelligence* (2007), *not 'free' at all.* It has to come from somewhere incredibly *intelligent.*

'Doesn't that point alone refute the materialist world view?'

'Yes, of course it does. What is so frustrating, Ali, is that this should be commonsense. All our experience teaches us that only *minds* and *intelligences* can create or *intelligently* evolve the highly and **specifically** (not randomly) complex information involved in the building and workings of *all* intelligent mechanisms be they children's toys, Jumbo Jets or tiny Living Cells "far more complex than Jumbo Jets" (as a molecular-biologist I know once said). The same goes for all CODES, be it Morse Code, computer languages like COBOL and BASIC or the DNA CODE.

'Obviously, in the case of our technologies the super or *higher* Natural causes of the mechanisms we make are *we* ourselves, who are, in classical spiritual understanding, the intelligent microcosms of the

---

[43] Gates, B. 1995. *The Road Ahead.* Viking.
[44] Gitt, W. 2006. *In the Beginning Was Information.* Master Books, p. 124. Meyer, S. 2009. *Signature in the Cell.* Harper One.

macrocosm, and in the case of the Life Forms it is other intelligent agencies that perform that function – subtle, superordinate, inter–dimensional laws, forces and agencies – agencies whose existence cannot be *physically* seen but can be *rationally inferred* and spiritually cognized, to some extent, by those with the developed psychic or extrasensory faculties to do so.

'Today, however, those classically minded idealists or intelligent design thinkers who challenge the materialist view in biology put their careers at serious risk. Disappointingly, Ali, I've slowly come to see that this topic, the origins of Life and its evolution – just as the question as to whether the Sun orbited the Earth or vice-versa, in Galileo's day – is one in which many of the usual laws of evidence-based scientific thinking are simply ignored or *even turned completely on their heads* because it is so overlaid with cultural subtexts and all kinds of social, emotional and political agendas that have nothing to do with classical scientific truth.'

'For example?'

'For example scientific frustration with dogmatic religious interpretations of reality that no longer make sense. The doctrine, for example, of one apparently random Life and then an eternity of a boring sounding 'heaven' or a truly hellish 'hell'.'

'But not all the world's spiritual traditions insist on such things.'

'That's true, many go for more sophisticated descriptions of holistic, multi–dimensional reality. Another agenda, though, is that in a culture where the religions and spiritual traditions are discredited and marginalized, it is the physical–scientists who get to be a new kind of 'high priesthood' and arbiters of the 'highest truths' at our culture's top tables. Once this new quasi high priesthood acknowledges the reality of the *intelligently* multi–dimensional and meaningful Cosmos in which we find ourselves it will no longer be seen to be quite so important.'

'Why though? Surely science will always be crucial?'

'Yes, of course, but it won't be physical–science alone, it will be the psychical and spiritual sciences as well. The Subjective, inter-dimensional worlds of Consciousness, Mind, Soul and Spirit will be understood, once more, to be self-existing principles in their own rights, not merely the allegedly 'accidental' and, ultimately, meaningless byproducts of equally meaningless and 'accidental' matter.

'But continuing to insist, Oliver, that we live within a mindless and mono-dimensional Cosmos of unmeaning and total accident seems to be a high price for the rest of us to pay to ensure the continued high status of a new materialist high priesthood which is no better than the old one – and possibly worse. At least the old high priesthood told us we had Souls and were the intelligent and meaningful offspring of Divine Cosmic Intelligence and Meaning not Soulless *'accidental chemical reactions', 'unconscious zombies' or 'nothing but a pack of accidental neurons'!*

'I agree, Ali. Thankfully, though, gradually more and more scientists are beginning to realize that the materialist idea that all the amazing, Soul Life Form Beings' clever, outer, chemical mechanisms required to originally arise, to evolve and to continue to arise, instant by instant, 24/7, 7/52, for millions of years, was 'free' Energy + 'free' Chemicals + 'free' Mindlessness + 'free' Unintelligence + 'free' Entropy + 'free' Accidents was as unrealistic as imagining that if we placed a transparent barrel in the Museum of Natural History in London or the Smithsonian in Washington, filled it with 'free' Watch parts or 'free' Car parts or 'free' Computer parts or millions of 'free' Scrabble letters or 'free' zeros and ones or 'free' mixtures of Oxygen, Carbon, Hydrogen, Nitrogen, Calcium, Phosphorus, Potassium, Sulphur, Sodium and Magnesium, and then caused the barrel to mindlessly, unintelligently and 'freely' tumble and tumble for a very long time – that when we opened it out would pop shiny new 'free' watches, 'free' cars, 'free' computers, the 'free' complete works of Shakespeare (or even just one line of Shakespeare), 'free' DNA CODING, "more advanced than any software ever created" or 'free' newly minted tiny Living Cells, "the most complex system(s) known to man".

'*Specifically* and intelligently complex information as conveyed by all codes and coding systems, like the digital DNA CODE inside Living Cells, is only *ever* known, as Stephen Meyer explains in his seminal book, *Signature in the Cell: DNA and the Evidence for Intelligent Design,* to arise from *intelligent* sources. As *even* Richard Dawkins *says:*

> 'What has happened is that genetics has become a branch of information technology. It is **pure information**. It's **digital** information. It's precisely the kind of information that can be translated digit for digit, byte for byte, into any other kind of information and then translated back again.[46]

'But surely that is precisely the kind of information that only ever arises as the result of intelligent input?'

'Yes, correct. As information theory expert, Professor Werner Gitt says:

---

[46] Richard Dawkins. 2008. Life: A Gene-Centric View, Craig Venter & Richard Dawkins: A Conversation in Munich. *The Edge.*

'The basic flaw of all [materialistic] evolutionary views is the origin of the *information* in living beings. It has never been shown that a CODING system and semantic [meaningful, intelligent] information could originate by itself ... The *information theorems predict that this will never be possible.* A purely material origin of life is thus [ruled out]'[47]

'Genetics and the DNA CODING, as even Dawkins confirms, without condeding the implications, is pure, *digital information,* designed to convey highly specific *instructions* which are carried out, 24/7, by the amazing, intra-cellular, bio-robot ribosomes at incredible, *supercomputer speeds*. And, as Dembski says, there are no 'free lunches' when it comes to codes and software. The specific (non-random) complexity that characterises all codes cannot arise by unintelligent and random means. That is not merely improbable, it is, as W. Gitt also points out, mathematically and statistically impossible.'

'So it is unreasonable to to continue with the methodologically materialist or (so-called) naturalist hypothesis that these amazingly clever processes are the products of random chance?'

'Yes, of course it is, and, just as some modern machines are programed with machine Codes to print out physical objects, the amazing ribosomes, following the digitally Coded instructions in the genomes, print out Life's proteins, at *astonishing* speeds, bringing the Soul-Life-Forms of the Plants, Animals and People into physical reality.

'This digital dna CODING conveys intelligent, instructional information for creating forms (in-form-ation) and can no more be thought of as an *unintelligently designed* chemical accident than COBOL, BASIC, Morse code or Mozart's music could be. This coding system whatever its subtle, inter-dimensional sources, whatever their precise nature, whether we see this as spooky or exciting, *can only be intelligent* in origin – amazingly so.

'Excitingly we are at a tipping point. An intellectual equivalent of the fall of the Berlin Wall is imminent. Like Sleeping Beauty we pricked our fingers on Darwin's materialistic pin 150 years ago and it sent us to sleep. Thankfully people are beginning to wake up. Life on Earth cannot be explained away anymore as a meaningless *'chemical accident'*. The light of classical idealist, intelligent design thinking is beginning to flow back in and to illuminate Nature for us instead of dumbing it down.

---

[47] Gitt, W. 2006. *In the Beginning Was Information.* Master Books, p. 124. Emphasis and text in square brackets added.

'You see, originally, the enterprise of science began because people, noticing their own *minds and intelligences* and the amazing *intelligences* embodied in the plants and animals and the orderly and predictable movements of the planets and stars assumed that these had *intelligent,* not mindless sources. After Darwin, however, science ironically became an activity which was to *intelligently* explain how *unintelligent* Nature is. Philosophical science began to sound like this:

> 'Science says, this is how gravity works, it has no 'purpose' or 'meaning' it's an accident. This is how electricity works and how we can use it, but, fundamentally, it's meaningless, it's an accident. This is how atoms work. Today we can make incredibly 'smart-materials' with them, that do amazing things, but they're nothing special, their apparently clever functionalities and teleologies are just meaningless coincidences. This is how Living Cells work, 24/7, they seem to be astonishingly clever, miraculous even, but, actually, it's an illusion, they have no purpose or teleology, because they're meaningless and accidental too.'

'This is sad, and is based on an intellectual misunderstanding. The misunderstanding that to describe how something **works** is a full explanation. It is, however, ***only a partial explanation***, it is but a description of materials and mechanisms. It is not an explanation of ideas and purposes, sometimes also referred to as formal and final causes. It doesn't tell us either that a particular mechanism or organism arose 'by accident' or that it continues to arise right now, second to second, 'by accident'. Saying that: 'The smart-materials-chemicals and the stunningly clever laws that govern their behavior are caused by a Big Bang' is but a description of an order of events. It is not a full explanation. It simply adds some technical elucidation to the description of reality provided by those who say: 'Higher Nature, God or Supreme Being did it'.

'Why? Isn't the Big Bang a good explanation?'

'The trouble is, Ali, it's not a full explanation because it simply involves 'matter' and 'mechanism' (material and mechanical causes). The second explanation, for its part, while it has no technical, mechanical or physical–science insights to offer, is an expression, however simple or advanced of spiritual–science, that is it, allows for meaning, shaping ideas and purposes (formal and final causes). It is only by bringing together these two sets of aristotelian causes (materials and mechanisms, the domains of physical science, and ideas and purposes, the domains of spiritual science, of which the world's religions and esoteric traditions are parts), that we'll ever achieve a more complete and Holistic explanation of Nature.

'Materialist thinking biology, however, because it does not believe Nature to be *purposeful and intelligent,* has to hypothesize that Living Cells, "the most complex systems known to man", are nothing more than a continuation of mineral chemistry 'by other means' even though it hasn't a clue as to what those other means might be, especially as it denies the intelligent Life and Soul Forces in all Living things, mostly, because someone artificially synthesized urea in 1828. Materialist thinking science also believes that to admit that there might be 'super' or Higher Natural components to Life is 'giving up' on science and giving up on methodological–naturalism, which is the idea that all things in Nature have purely 'ordinary' Natural explanations and never 'super' Natural ones.

'But what is 'unnatural' about 'higher' or 'super' Nature, Ollie? If it *exists* and is what drives Life, by whatever subtle laws, forces and processes, which we do not yet fully understand, then it is as much a part of Nature as anything else and can hardly be called 'unnatural'.'

'I agree, Ali. Our culture makes these rigid distinctions between the so-called 'ordinary' Natural and the 'super' Natural as though, somehow, they are completely separate when, really, they are all one. Personally, I don't mind what labels we use but the materialist or so called naturalist dogma that it is the task of science to explain all things in 'natural' terms, where 'natural' means 'mindless, unintelligent and accidental', leads to intellectual absurdities. For example methodologically materialist biology's unintelligent design explanation of the ***digital information*** in Cells, the DNA CODING enigma goes something like this:

> 'We know from our universal experience that CODING systems always necessitate the input of intelligent information and that they only ever have intelligent sources. However, the DNA CODING, "like a computer program but far, far more advanced than any software we have ever created,"[48] is a unique exception. It, unlike other coding systems, 'has to be' an 'accident' of some kind, because the materialist metaphysic of a meaningless and accidental Cosmos requires it.
>
> This means that in defiance of the usual laws of entropy and of probability theory, not to mention common sense, which make 'accidental' CODES not merely improbable but impossible,[49] the DNA CODING 'accidentally wrote itself'.

---

[48] Gates, B. 1995. *The Road Ahead.* Viking.
[49] Gitt, W. 2006. *In the Beginning Was Information.* Master Books, p. 124. Meyer, S. 2009. *Signature in the Cell.* Harper One.

'Very funny, Ollie.'

'You say that but those statements do flow from the materialist world view which refuses to take any notice of Aristotle's famous last two causes, formal and final, that is intelligent shaping forces and purposes and believes in a kind of unquestioning way that it can successfully explain all of Nature's mechanisms in terms of Aristotle's first two causes alone, *just materials and mechanisms,* which, without his last two causes, always have to be 'accidents'.

'Hence 'accidental' quantum fields, accidental Big Bang, accidental light, accidental electricity, accidental gravity, accidental atoms, accidental molecules, accidental digital DNA CODING, "more advanced than any software ever created", accidental cells, "far more complicated than any machinery made by man", accidental plants, accidental people, accidental planets, accidental stars, accidental everything.

'Personally, I think the excessively materialist posture of much of modern science, for which I otherwise have enormous admiration, is philosophically unreasonable because if Nature really was 'unintelligent' we would not be here to *intelligently* explain it, it wouldn't be *intelligently explainable* and nothing would have any true meaning or make any kind of sense at all which, of course, is the last stop on the topsy turvy, materialist line – mindless, zombie unconsciousness.'

'But surely the materialist or so called naturalist difficulty, Ollie, is that how could intelligent laws, forces, processes or agencies capable of giving rise to Living Nature or indeed the entire universe exist? How could they exist at all? Where would they come from? Who'd make them? Who'd make the 'intelligent designers' of the intelligent and astonishingly precise Natural laws and processes?'

'Yes, that is the materialist difficulty but, in essence, we have *only two ways* to answer that question or the riddle of the ultimate Mystery of Existence – why anything exists at all as opposed to nothing at all, not even one photon, and as to whether what does exist, the Universe, our Living World and our own Minds and Intelligences, for example, is sourced in Mindlessness, Unintelligence and Accident or MUA or in purposeful intelligence and amazing, ultimate meaning or AUM which is the Idealist view, currently the minority view in science and biology.'

'Okay ...'

'All we can do is decide what *kind* of Cosmic Mystery of Existence we believe we are in – Is it a mindless and accidental Mystery of Existence or an intelligent and meaningful Mystery of Existence? We cannot, however, solve the mystery of its Final Source and *why* it Exists at all.'

'How do you mean?'

'Well, the materialist hypothesis of Nature, of total Reality is that 'accidental matter' *just exists* (who knows why or how?) or that there was a Big Bang (who knows why or how?) that 'accidentally' gave rise to the material universe and that, by an endless series of meaningless coincidences, this eventually led to Life and Consciousness and Mind and subjectivity on this planet and possibly elsewhere. The materialist view does not answer, however, *why* incredibly smart-materials–chemical 'matter' (made of intelligently informed quantum energy s—p—a—c—e), and governed by incredibly intelligent laws, should just exist or appear in a Big Bang, from nowhere to now here, all it can do is say: *"It just does!"*

The materialist view sometimes claims, with far less evidence than there is for the 'super' Natural and the *intelligently causal* multi–dimensions of total Holistic Natural Reality of classical idealist inference, religious belief and direct psychic and spiritual experience, that if (and there's *no empirical evidence* for this at all) universes or multiverses are forming more or less every second, some of them are bound to contain laws like those in our universe that 'seem' intelligent. This, however, does not work, because in that kind of multiverse, in which anything is possible, then, this being a logical possibility, a blazingly intelligent Final Source or Uncaused Cause or Supreme Beingness must logically be there too and doing what Supreme Intelligent Beingness does, which is give rise to All That Is, including however many universes you like.

'In line with this latter proposal, the classical idealist, ('mind-before-matter-and-mechanism') view says that self–existing, self–arising Beingness containing potentials of blazing power and intelligence *just exists,* (who knows why or how? – the Final Source and Uncaused Cause to which people give names like Supreme Being, God, the Tao, Allah, Brahman etc), that it gives rise to the apparently solid, 'material' universe (made, ironically, of nothing but intelligently informed quantum s—p—a—c—e) and that it expresses itself as all dimensions of reality, hologrammatically Self-fragmenting into trillions of existences with both inner and outer aspects, subjective, mental and intelligent and objective mechanical and smart, including our Living World. Neither the classical Idealist nor the traditional religious and spiritual world views can say, however, *why* an amazing, self–arising, self–existing, Final Source and Uncaused Causing of dazzling intelligence and stupendous Cosmic–scale potentials should *just exist* or manifest in a Big Bang, from nowhere to now here, all they can do is say: *"It just does!"*

'So you see that we have these two conflicting visions of total Holistic Nature. The materialist view is comfortable with 'solid' 'm—

a—t—t—e – r' (made of quantum s—p—a—c—e) *just existing,* who knows why, but it is entirely uncomfortable with the idea of intelligent Consciousness, Creativity, Life, Soul and Spirit *just existing,* who knows *why,* and intelligently and meaningfully giving rise to 'All that so amazingly Is' including our Living World and our own universal witnessing, observing consciousnesses, minds and intelligences. Materialism while comfortable with measurement and quantity, seems to really dislike the idea of *intrinsic intelligence,* of Soul qualities, of true Life, and Mind and meaning and denies that the Universe contains any of these, ironically using, all the while', its own immaterial qualities of *mind and intelligence* and ability to make meaning to do so.

'So we can't escape the Mystery, Ollie, as to *why or even how* there should be anything at all rather than nothing at all? Be it:

(a) just mindless, unintelligent matter plus an infinity of accidents as Materialism holds or

(b) self-arising, self-existing, Consciousness, Intelligence, Purpose and Meaning, giving rise to a vast extending expression of numerous levels, dimensions, vibrations and frequencies of Minds, Souls, Intelligences, materials and mechanisms all working holistically in synergistically nested hierarchies of stunningly dynamic 24/7 functioning, as Idealism holds?'

'Exactly. The materialist view accepts that mindless matter 'just exists' or that it came from a Big Bang, for no rational reason that materialism can come up with, other than: "Well, it's just a vast MUA! It's mindless, unintelligent and accidental. *That's just the way it is!"* That is the materialist summary of existence. It is the one that dominates science today, paradoxically, even bio-logy the science of the most 'super' Natural and 'miraculous' phenomenon of all, very Life.

'The philosophically idealist view, on the other hand, currently, the minority scientific view, accepts that consciousness, intelligence, purpose, love, beauty all *'just exist'* and give rise to all that 'ordinary' Naturally and 'super' Naturally is for no rational reason that philosophical idealism, or the world's religions and spiritual traditions, which are also idealist in philosophy, can come up with other than: *"That is just the way it is!"* Existence is, in the idealist view, an intelligent and purposeful affair, an amazing, ultimate, miracle or an AUM. This is the idealist and classical religious and spiritualist summary of existence.'

'So all we can do, Ollie, is decide which kind of mystery we believe we are in and whether an *intelligent and meaningful* Mystery of existence makes better sense of the data of Nature and of the Living World or a *mindless and accidental* Mystery of existence makes better sense of the data of Nature?'

'Yes, that's what I believe.'

'But the answer seems obvious to me, Ollie.' exclaimed Alisha.

'I agree, I think it is too but as always with humanity the picture is complicated by prejudices and social and psychological factors.'

'Such as?

'Such as the dislike, in some parts of the scientific community, of the organized religions for example. If we can keep the idea going that the Living World is all one big 'accident', it definitely makes the world's religions meaningless and their officials more or less redundant, doesn't it?'

'But why would anyone want to do that?'

'The religions can be pretty bossy, can't they, Ali? Historically they have, at times, held back progress and have not always been forces for enlightenment. And, even today, in countries where they wield a lot of political influence, they can be really quite oppressive at times.'

'I see what you mean.'

'But don't get me wrong, I don't think that is a good enough reason to deny the Living World's intelligence. The trouble is the ideological–materialism that dominates science and biology today can never admit that Nature is purposeful and intelligent because this *refutes and falsifies* the materialist world view.

'The only way to avoid this gloomy conclusion to materialist thought, for surely it should be a happy one, is to carry on with the statistically impossible proposal that the digitized genetic information conveyed by the DNA CODING, "more advanced than any software ever created" and the intellectual distortion, or delusion, that tiny Living Cells "more complex than entire factory-cities, in spaces smaller than sand grains" are *'unintelligent' and 'purposeless' phenomena – they are 'accidents',* as, indeed, are all the Life Forms right up to and including the minds and works of Leonardo da Vinci, Vivaldi, Mozart, Picasso and Einstein. If that's not a profoundly unreasonable and anti-empirical, even dishonest ideology, Alisha, I don't know what is.'

'That's putting it a bit strongly, Ollie, isn't it? Materialist thinkers are surely, just like everyone else, simply arguing for what they believe to be true.'

'Yes, but what is so depressing about materialist thinking is that in order for its dark and gloomy world view to seem intellectually respectable (a) it has not only to make science false to reality and our language meaningless by claiming that Living Nature is unintelligent, non–teleological and that 'accidents of chemistry' explain all, Life, Mind, Intelligence, Love, artistic and mathematical ability etc when there is no real evidence to support that hypothesis and a great deal to falsify it, but also (b), worse than that, it has to actively and skeptically debunk and deny *so many areas* of real, deeply lived and deeply felt human experience, including the most precious and *highest* experiences any of us can have, for example our mystical, peak, religious and spiritual experiences of all kinds,[50] pointing to the true, multi–dimensional nature of reality, that we really do have Souls and that we really do survive the 'death' of the chemical body.[51] Materialism, however, denies that all such revelations, insights and experiences are anything but 'accidents of brain chemistry', denying that they have any objective truth or public knowledge-value.'

'But are people's private psychic and spiritual experiences of any public relevance?'

'Yes, of course they are. For example, some people have died, during operations, and found themselves outside their bodies, then drawn through a tunnel of Light to the next dimension or vibration of holistic existence where they were greeted by loved ones already passed on. Eventually, they were drawn back, often somewhat reluctantly, into their bodies and physically revived. There are many such accounts and there is, today, a whole literature and research relating to this phenomenon called NDEs or Near Death Experiences:[52]

> 'A college professor … told of dying on the operating table and then instantaneously finding himself walking down the gentle slope of a bright greensward. No one else was there. Nothing else happened. Yet to hear him speak of it, it was as if the greatest of miracles had happened to him, and he now knew, he absolutely knew, that there was life after death. This brief incident completely transformed his life. Even today he lights up describing the incredible aliveness of the green grass he once walked upon on The Other Side of death.

---

[50] James, W. 1902. *The Varieties of Religious Experience.* Penguin Classics. Bucke, R. M. 1905. *Cosmic Consciousness: A Study in the Evolution of the Human Mind.* Innes and Sons. Kelly, E. F. & E. W. Kelly, A. Crabtree, A. Gauld 2009. *Irreducible Mind: Toward a Psychology for the 21st Century.*

[51] Fontana, D. 2005. *Is There an Afterlife?: A Comprehensive Overview of the Evidence.* O Books. Carter, C. 2010. *Science and the Near Death Experience.* Inner Traditions.

[52] For example: Lommel, P. 2010. *Consciousness Beyond Life, The Science of the Near Death Experience.* Harper One. Fenwick, P. & E. Fenwick. 1995. *The Truth in the Light: An Investigation of Over 300 Near Death Experiences.* Headline Book Publishing.

'I have spoken with many people who have described 'the living dark' that [first] greeted them with words like 'soft velvety blackness' and 'warm inviting blanket'. ... these people felt awed by the wonderment of a blackness that appeared *intelligent,* emoted feelings and instilled in the experiencer a sense of peace and acceptance. ...

'[Others] encountered light... They claim that the radiant brilliance of this special light does not blind or burn; it simply accepts, embraces and loves. A sense of worthiness can remain afterwards, changing forever how the experiencer regards him or herself.'[53]

'These are individual experiences but they are of enormous interest to all of us, surely?'
'Why, though?'
'Because they show that consciousness and awareness are not, ultimately, of the body, they are not 'material', they point to the existence of Life after death, that our conscious selves *do* survive the death of our chemical bodies, as collective humanity has always intuited and believed and the world's religions and spiritual traditions have taught. Often those who have had these NDEs are greatly changed by them. They are more at peace and serene and they have lost all their fear of death.[54]
'Surely it's a great gain for our culture to be aware this?'
'I suppose it would make a lot of people less afraid of death and dying if they really believed death was not the end. But are these people really dead?'
'Yes, in many cases they are certifiably, clinically dead – that is well and truly dead, by all normal and exceptional definitions of the term. However, the materialist thinking, that dominates science and biology today, is simply uninterested in such accounts.'
'Why though?'
'Because *they refute the materialist paradigm.* Methodological–materialism is not a scientific approach with an open mind, Ali – it's just a world view that decided long ago that Life and existence are purposeless and unintelligent phenomena, they are 'accidents', it rejects any and all evidence that suggests otherwise and it maintains that its particular world view and science are one and the same which they most emphatically are not. This is the same mindset that rejects all the obvious evidence for intelligent design and *intelligent* evolution in the Living World.'

---

[53] Atwater, P. M. H. 1994. *Beyond the Light: Near Death Experiences.* Thorsons; p. 28.
[54] Grey, M. 1985. *Return From Death: An Exploration of the Near Death Experience.*

'But surely, Ollie, there is, nonetheless, a difference between objective and subjective experiences?'

'It depends on what you mean by objective. Look, a tree is not made objectively true just because we can all see it. It is objectively true because it has its own Being, its own essence, it exists in its own right. The seeing of the tree is *subjective* in all of us even if it is a seeing or an experience that most of us can corroborate. This is what, many people think, allows us to call it objective even though we will all experience the objective tree slightly differently.

'Materialist scientific thinking endorses our collective *subjective* experiences of the physical tree. It is not prepared, however, to endorse the truth of numerous individual subjective experiences, such as our psychic, religious, mystical, peak and spiritual experiences, out of body experiences and near death experiences, OBEs and NDEs, *even when they share many commonalities* and can often be corroborated in various ways, as can be seen in the modern research into these phenomena which share many similarities.[55]

'In materialist *thought* (which is, like all thinking, a *subjective* process – that is it needs a Subject, the mysterious, Conscious, universal witnessing awareness and it's accompanying human *Intelligence* to undertake it) these experiences are all reduced to meaningless anecdote. We think of calling someone at precisely the same time as they call us but in materialism this is just a meaningless coincidence, not a meaningful one.[56] Two or more of us share a spiritual experience, perhaps the special or peculiar atmosphere of a place we visit and which we all notice, but materialism dismisses our experience as meaningless. Many people witness a psychic or paranormal phenomenon of some kind, like the famous Fatima revelations in Portugal, in 1917, but, again, materialist thinking refuses to acknowledge the objective and empirically evidential nature of the experiences because they are not ones that can be physically weighed or measured or transferred to a lab.

'This, though, is the same materialist thinking that is entirely untroubled by the fact that *'accidental chemical reactions'* never

---

[55] Carter, C. 2010. *Science and the Near Death Experience.* Inner Traditions. Gustus, S. 2010. *Less Incomplete: A Guide to Experiencing the Human Condition Beyond the Physical Body.* O Books.

[56] Rupert Sheldrake: "Most people say they have experienced telepathy, especially in connection with telephone calls." Sheldrake has conducted "tests to find out if people really could tell who was calling them on the telephone when the caller was selected at random. The results were far above the chance level." sheldrake.org.

build 'working machinery' of any kind, entirely untroubled by the fact that there is no laboratory evidence that Life is a chemical accident – the famous Miller–Urey and all its successor experiments included[57] – or that digital DNA CODING could never arise 'by random chance' processes – it's not merely improbable, it's impossible – any more than any of our far less sophisticated CODING systems ever do.[58]

'No one's experience, even of a tree which we can all see, is truly accessible to anyone else. What is so unreasonable about the materialist perspective on existence, Ali, is that in normal reality, if just two or three people witness a crime, it is enough to convict the criminal. However, even if *countless people,* all over the world, for centuries, testify, as they have done and continue to do, to subtle mystical, psychic, inter–dimensional or spiritual experiences, which are as real to their psychic or spiritual senses as physical trees are to our physical senses, then their experiences are dismissed, in ideological–materialism as of no public knowledge value at all.[59]

'Such experiences, while they cannot be corroborated nearly to the same extent as our experience of a tree, can, however, often still be corroborated. For example, the many commonalities in people's out of body and near death experiences as they have been recorded by researchers, over the years. It is only a strange and almost perverse refusal in our scientific and philosophical culture to apply Occam's razor, that is to apply the most parsimonious explanation to these experiences, which is to accept that they are what they appear to be, evidence that minds and consciousnesses can exist apart from bodies, which causes the forces of materialist thinking to continue to hypothesize that, while real to those having them, they have no objective content or truth value.'

'The problem is that, as Chris Carter says in his extremely thorough evaluation of modern near death research, *Science and the Near Death Experience: How Consciousness Survives Death,* the ideological materialism that has captured modern science and biology "simply cannot accommodate corroborated reports"[60] that human consciousness can operate outside of a physical body. Many scientists *simply refuse* to accept the great weight of modern evidence that falsifies the

---

[57] Wells, J. 2002. *Icons of Evolution.* Regnery; pp. 81–111.
[58] Meyer, S. 2009. *Signature in the Cell.* Harper One.
[59] Radin, D. 2009. *The Conscious Universe: The Scientific Truth of Psychic Phenomena.* Harper One.
[60] Carter, C. 2010. *Science and the Near Death Experience.* Inner Traditions; p. 235.

materialist world view including that for OBEs, NDEs, messages from those passed on[61] and for reincarnation.[62]

'As Chris Carter says: "When many scientists and philosophers are confronted with the evidence, *their reaction is often anything but rational.* Philosopher Neil Grossman describes how he discovered this for himself:"

> 'I was devouring everything on the near-death experience I could get my hands on, and eager to share what I was discovering with colleagues. It was unbelievable to me how dismissive they were of *the evidence.* "Drug-induced hallucinations," "last gasp of a dying brain," and "people see what they want to see" were some of the commonly used phrases. One conversation in particular caused me to see more clearly the *fundamental irrationality of academics* with respect to the evidence against materialism.
>
> 'I asked, "What about people who accurately report the details of their operation?"
>
> '"Oh," came the reply, "they probably just subconsciously heard the conversation in the operating room, and their brain subconsciously transposed the audio information into a visual format."
>
> '"Well," I responded, "what about the cases where people report *veridical perception of events remote from their body?"*
>
> '"Oh, that's just a coincidence or a lucky guess."
>
> 'Exasperated, I asked, "What will it take, short of having a near-death experience yourself, to convince you that it's real?"
>
> 'Very nonchalantly, without batting an eye, the response was: "Even if I were to have a near-death experience myself, I would conclude that I was hallucinating, rather than believe that my mind can exist independently of my brain."'[63]

'Materialism is a belief system, a specific faith position. As Carter says: "For such individuals, *materialism is not a scientific hypothesis that is open to potentially being proved false; it is **an article of faith** that "must" be true,* regardless of evidence to the contrary." It is the conflation of the ideology of materialism with the methodology of science that causes so much confusion. Chris Carter explains that it is the:

---

[61] Schwartz., G. E. Phd. & W. L. Simon. 2002. *The Afterlife Experiments: Breakthrough Scientific Evidence of Life After Death*. Atria Books.

[62] Stevenson, I. 1980. *Twenty Cases Suggestive of Reincarnation: Second Edition, Revised and Enlarged*. University of Virginia Press. Stemman, R. 2012. *The Big Book of Reincarnation: Examining the Evidence that We Have All Lived Before.* Hierophant Publishing.

[63] Grossman, N. 2002. *Who's Afraid of Life After Death?* p. 8

'*implicit equation of materialism with science*, that explains the widespread practice of ignoring and dismissing the objectionable evidence as somehow 'unscientific'. Materialism is upheld as an incontestable dogma on which, it is thought, rests the entire edifice of science. But **the difference between science and ideology is** not that they are based on different dogmas; rather, it is that scientific beliefs are **not** held as dogmas, but **are open to testing and hence possible rejection**. Science cannot be an objective process of discovery if it is wedded to **a metaphysical belief that is accepted without question** and that leads to the exclusion of certain lines of evidence on the grounds that these lines of evidence contradict the metaphysical belief.'[64]

'Carter quotes Grossman again:

'Science is a methodological process of discovering truths about reality. Insofar as science is an objective process of discovery, it is, and must be, metaphysically neutral. Insofar as science is not metaphysically neutral, but instead weds itself to a particular metaphysical theory, such as materialism, **it cannot be an objective process for discovery**. There is much confusion on this point, because many people equate science with **materialist metaphysics**, and phenomena that fall outside the scope of such metaphysics, and hence cannot be explained in physical terms, are called 'unscientific'. This is a most unfortunate usage of the term. For if souls and spirits are in fact a part of reality, and science is conceived epistemologically as a system of investigation of reality, then there is no reason why science cannot devise appropriate methods to investigate souls and spirits'[65] [— As, in fact, a courageous minority of scientists has been doing for more than 150 years now.[66]]

'Contrast these views with the ideologically materialist perspective that is put so clearly by a well known biologist, Richard Lewontin. He says:

'Our willingness to accept scientific claims that are against common sense is the key to an understanding of the *real struggle*

---

[64] Carter, C. 2010. *Science and the Near Death Experience.* Inner Traditions; p. 237, emphasis added.
[65] Grossman, N. 2002. *Who's Afraid of Life After Death?* p.10–12; emphasis added.
[66] Fontana, D. 2005. *Is There an Afterlife?: A Comprehensive Overview of the Evidence.* Schwartz, G. E. Phd. & W. L. Simon. 2002. *The Afterlife Experiments: Breakthrough Scientific Evidence of Life After Death.* Carter, C. 2010. *Science and the Near Death Experience.*

between [materialist] science and the supernatural. We take the side of [materialist] science in spite of the patent absurdity of some of its constructs... in spite of the tolerance of the scientific community for *unsubstantiated* just so stories, because we have ***a prior commitment,*** a commitment ***to materialism.*** It is ***not*** that the methods and institutions of science somehow compel us to accept a *material* explanation of the phenomenal world, but, on the contrary, that we are *forced* by our a *priori* adherence to material causes [alone] to create an apparatus of investigation and a set of concepts that produce *material explanations,* no matter how counterintuitive, no matter how mystifying to the uninitiated [those who do not think materialistically]. Moreover, that ***materialism is absolute***, for we cannot allow a Divine foot, [or *intelligent* causation], in the door. The eminent Kant scholar Lewis Beck used to say that anyone who could believe in God ['higher' Nature, *intelligent* causation in Nature] could believe in anything.'[67]

'Lewontin expresses the view that science *is* methodological–materialism and that it is a positively 'anti-supernatural' activity – that science is, literally, in a struggle with the 'super' or 'higher' Natural and the intelligent and meaningful multi-dimensionality of existence. This view assumes and insists that total Nature is simply a vast accident. That is very different from a scientific hypothesis of materialism that is *open to falsification,* open to being refuted. Ideological–materialism narrows science down, it narrows down our understanding of reality and our potential to understand reality. Methodological–materialism is a science blocker. Key parts of mainstream science, today, have been distorted by the ideology of materialism for it is not metaphysically neutral. This is why lines of evidence that contradict materialism as a metaphysical theory are so strongly resisted.

'Because, Ollie, once such lines of evidence are admitted materialism collapses as a scientific hypothesis?'

'Yes, exactly. In fact, there are already many lines of evidence that have falsified materialism as a scientific hypothesis.'

'So what's the problem? Is it not 'job done' then?'

'The problem is that science is a social and political activity and 'true' in science is not necessarily what is actually true but what the majority say is 'true'. So, if, in spite of the evidence, the majority continue to maintain that such lines of evidence have not falsified

---

[67] Richard Lewontin (1997) Billions and billions of demons (review of *The Demon-Haunted World: Science as a Candle in the Dark* by Carl Sagan, 1997). *The New York Review*, January 9, p. 31. Emphasis and text in square brackets added.

materialism, then materialism will continue to be 'true' in the same way that it continued to be 'true', for a time, in Galileo's day, that the Sun orbited the Earth even though he had shown that it did not. Carter rightly says that:

> 'the term "scientific" carries a lot of emotional weight. In our modern world, science and scientists hold a great deal of prestige, and so few people want to be thought of as unscientific. To be labeled unscientific is enough to have one's work dismissed from serious consideration by the academic establishment. If to be scientific is good and unscientific bad, and if the term "scientific" is thought to be synonymous with the term "materialistic," then any talk of disembodied minds or spirits is anti-materialist, unscientific, and therefore bad.[68] ***The longstanding confusion of materialism with science*** is what largely accounts for the persistent social taboo responsible for the ignorance and dismissal of the ***substantial amount of evidence that proves materialism false.***'[69]

'Carter quotes Bruce Greyson, a world renowned expert in the field of near-death studies, who says of the damaging conflation of science with the ideology of materialism that:

> "Materialists often claim credit for the scientific advances of the past few centuries. But ***it is the scientific method of empirical hypothesis testing,*** rather than a materialistic philosophy, that has been responsible for the success of science in explaining the world. If it comes to a choice between empirical method and the materialistic worldview, the true scientist will choose the former"'[70]

'The trouble is, Ali, too many today, even very well known scientists, choose the materialist worldview over and above the empirical method. This is an intellectual betrayal. A betrayal of the trust that we the public put in science as a source of genuine knowledge in our culture.

---

[68] Carter: 'For the materialist, the term unscientific seems to be the modern equivalent of the term heretical, and it is invoked for the same purpose: to exclude from consideration ideas that challenge the believer's faith.'
[69] Carter, C. 2010. *Science and the Near Death Experience.* Inner Traditions; p. 238.
[70] Greyson, B.: 'Commentary on "Psychophysiological and Cultural correlates Undermining a Survivalist Interpretation of Near-Death Experiences."' *Journal of Near Death Studies* 26, no. 2 (Winter 2007); p. 142.

We continued on in silence for a few moments, by the side of river, the meadow grasses brushing our feet.

'The problem is that for materialist thinking science's gloomy vision of Life and the Living World as accidental and meaningless to seem true, it really does have to deny or ignore all the evidence, such as that collected by Chris Carter and many other contemporary thinkers and scientists, that collective humanity's nearly universal intuitions that Life is something more than 'matter' and that its ultimate sources are *intelligent*, (not meaningless and 'accidental'), is correct. Materialism is not a value neutral philosophy of science. Far from it. It is positively anti-spiritual, positively anti 'super' naturalist. It has to be. This is because if the spiritual elements and dimensions of total Nature are real, then materialism, as a world-view, is shown to be false.'

'But wouldn't people be glad to realise that materialism, as a scientific hypothesis of reality, is false, that Life and Mind are, after all, more than mere physics and chemistry, let alone 'accidents'?

'You'd think so. But, strangely, although it is unreasonable, even irrational, in so many respects, a lot of thinkers seem to prefer the materialist ideology of mindlessness and accidental chemistry to explain all to the classical multi-dimensional or spiritual view that Life and Mind and the Living World are all stunningly clever expressions of purpose and intelligence, no matter how mysterious and difficult to comprehend, the ultimate sources of such purpose and intelligence may be.'

5 Materialist–Science: Tries to Abolish the Soul and Life After Death

Is it really true that sacred, amazing, miraculous Life is, as materialism holds, just accidental 'dead' chemistry? And was the artificial synthesis of urea in 1828 good evidence, really, for the materialist hypothesis that there are no Vital Forces in Nature or Soul or Life after Death?

We stopped walking and enjoyed the beautiful scene. The green valley, the sound of the water in the sparking river. The murmuring of the cattle and calling of the lambs.

'Look, there's a heron fishing. It's so stately.'

'Yes, it has such a powerful presence.'

'And over there ... on the other side of the valley, the deer moving up the hill, towards the woods.'

We sat down at the foot of a Sweet Chestnut Tree, close to the river, and opened our flasks. After a few moments quietly sitting and sipping our hot drinks we began our conversation again.

'Oliver, do you think anyone really thinks that the heron, these sheep peacefully grazing here, the birds flitting to and fro, these delightful meadow grasses are really all just second by second *'accidental chemical reactions'?* The amazingly intelligent effects of utterly mindless and unintelligent causes?'

'I hope not – but that is the stated position of materialist thought.'

'But it just seems so ludicrous and unreasonable ...'

'I agree Ali. But I think it has so much to do with the simplistic, medieval conceptualizations of intelligent causation in Living Nature that collapsed with the Victorian realization that Life on Earth was millions of years old, not thousands, and that the Universe could not possibly have been created instantly, in six busy days, six thousand years ago by a bearded-god type being.

'The advanced, yogi–adept, spiritual–scientists of India, however, unlike the Victorian materialists, were not burdened in this way because they'd been saying, for centuries, that the Universe and the Earth were billions of years old, not thousands. And, of course, had this knowledge been part of our western religious and spiritual tradition perhaps the Darwinian interpretation of the Fossil Records would have been very different.'

'Why?'

'Because the advanced yogis not only maintained that Earth was extremely ancient but that vast cycles of Life or Yugas, which lasted for millions of years, took place as the Life Forces waxed and waned, with times of expansion and Life Force revival and times of dissolution and decay. It is possible that these yogic yugas may, in fact, correspond with the patterns of Life Form renewals, long periods of stability or equilibrium, followed by periods of destruction and disappearances seen in the Fossil Records, the form of evolution that the late S.J. Gould called punctuated equilibrium. For a discussion of this intriguing idea it is worth reading John Davidson's *Natural Creation or Natural Selection?*[71]

'The Victorian materialists, however, were not aware of the yogic understanding of creation, which was that an *intelligent,* cosmic scale, eons-long, *Creation-in-Evolution* or ELCIE was underway and interpreted the new discovery (for them, not the yogis), of the extremely ancient age of the Earth very differently. For them it was 'proof' that probably the world's religions and spiritual traditions were all just wishful thinking, rather than simple or more advanced forms of spiritual–science, genuine mystical insights and spiritual revelations, however partial, into the 'super' or higher Natural causes of things, the spiritual, multi–dimensions of existence. They began to doubt that anything other than chemical matter was real.'

'In other words they began to be *materialistic* in their thinking?'

'Exactly. They gave up on classical, Idealist ('mind-before-matter-and-mechanism') dualism and became materialist monists, seeing reality as *all* 'accidental' matter, *all* 'accidental' object, *all* 'accidental machine', with no ghost anywhere to be seen, all tails with no heads or Subjective dimensions, no universal Cosmic Witnessing Awareness to create and collapse the cosmic quantum-wave function and bring all into existence. They began to see Nature as all ghostless, 'accidental machine' with *no truly* Intelligent and Subjective Mind or Soul components to it at all.

'Wasn't that an overreaction to their discovery that the Earth was millions of years old, not thousands, and had not literally been made in six days out of spit and clay?'

'Yes, I think it was. But for the Victorian materialists that is how they seemed to think. Either the world's scriptures had to be true in very literalistic ways (not just as spiritual knowledge but as physical–science too) or the *intelligent* spiritual causes of things were dismissed alto-

---

[71] Davidson, J. 1992. *Natural Creation or Natural Selection?* Element Books.

gether. And this is what happened. The result was a great loss of faith or belief in the Spiritual, in Life after death, in the Life Forces and in the Soul, and in *a complete reversal in western thinking*, which since the Greek times of Pythagoras, Socrates, Plato and Aristotle had been, predominantly, an Idealist ('mind-and-intelligence-before-matter-and-mechanism') intelligent design or ID tradition, to a non-evidence-based, materialist ('unintelligence-and-accidents-create-everything-clever-in-Nature') unintelligent design or UD tradition.

'Many people stopped believing in the independent existence of the Soul, apart from the chemical body, and in Life after death. There was also the all-important loss of belief in any kind of **Life Forces** or Vital Forces which leant support to the materialist notion that magical, sacred, miraculous Life itself is not *a self-existing principle* but is really just a weird form of accidental, Life–Forceless, chemistry – in other words, it's not really *Life* at all, just 'dead' chemicals doing their second to second, purposeless, zombie, *accidental-chemical-reaction* thing, for millions of years, for no rhyme or reason.

'But that's unreal, Ollie. Moment by moment, 24/7 bio-logical, *Living* complexity surely means that there is something more going on, *right now and now and now,* all day, every day, and for millions of years previously, than just *'accidental chemical reactions'*.'

'Of course. But in Darwinian materialism[72] nothing in Nature is *ever* allowed to be *intelligent* or to have a purpose or teleology because existence is an 'accident', consciousness is an 'accident', the universal witnessing awareness is an 'accident', creativity is an 'accident', subjectivity is an 'accident', purpose is an 'accident', intelligence is an 'accident', love is an 'accident', beauty is an 'accident', meaning is an 'accident', gravity is an 'accident', electricity is an 'accident', light is an 'accident', atoms are 'accidents', the magical, alchemical, smart–materials–chemicals are 'accidents', the digital DNA CODE, "more advanced than any software ever created", is an 'accident', tiny Living Cells, like 24/7 micro-miniaturized 'chemical-factory-cities', are 'accidents', everything is an 'accident'. Lions are 'accidents', elephants are 'accidents', lilacs and roses, clementines and limes, radishes, carrots, parsley and lettuce, the heather and the gorse, the mighty whale, the great elk, the soaring eagle and even all the people are all 'accidents' – none of them ever really truly, vitally, Soulfully alive because materialism declares that there *is* no Life Principle, there *are*

---

[72] Note: Not all Darwinians are materialists, some are theists, a peculiar position because the whole point of Darwinism is that it is an attempt at a no-intelligence-needed-or-involved explanation of the Living World.

no Vital Forces, there *are* no Soul forces, there is no true Consciousness, there are no driving intelligences or purposes in Living Nature. There are just dead, purposeless, 'accidental chemicals' having 'chemical accidents' and doing their pointless, millisecond to millisecond, 'accidental' thing.'

'How could they think that though? Surely it's obvious that stunningly complex, *dynamically alive,* Plants, Animals and People are all more than 'just chemicals' and that the words *"Life!" and "Alive!"* mean something more than 'just chemistry'. After all, that's why we call studying them *Bio*-logy rather than physics!'

'It's obvious to you and I, Ali, and any child who has ever lost a pet guinea pig or cat or beloved human relative, but to those Victorian, materialist thinkers it wasn't obvious at all. And the clincher for them was the fact that someone had managed, in 1828, to artificially synthesize Urea.'

'What has Urea got to do with the Soul or the Life and Vital Forces?'

'I agree, it hasn't or shouldn't have anything to do with them. But the original idea was that the reality of the Life or Vital Forces was demonstrated by the differences between *inorganic* and *organic* chemicals. Urea was an organic chemical as opposed to say oxygen or hydrogen, which are inorganic ones.'

'People, at the time, believed that organic chemicals were fundamentally different from inorganic chemicals and they took this presumed difference to prove the existence of the Life Forces. Urea, however, is an organic chemical, so when someone organ-ized it, that is artificially synthesized it out of in-organic chemicals the theory collapsed because people realized that, at base, organic and inorganic chemicals are *the same,* there's no difference between them anymore than the metals and plastics in TVs are fundamentally different from the metallic ores and crude oils in the ground. Our bodies are made, predominantly, of organic chemical compounds of the atomic elements oxygen (65%), carbon (18%), hydrogen (10%) and nitrogen (3%). But oxygen, carbon, hydrogen and nitrogen, in their pure states, are all inorganic chemicals, just like all the other chemical elements.

'The consequences of this decision about the Vital or Life Forces were truly dramatic. Once people began to think that there were no intelligently in-form-ing Life Forces, then it was but a short step to deciding that there probably weren't any Souls either – and that meant there was *nothing to survive* the death of the body, so no Life after death. These nineteenth century materialist thinkers were beginning to think that Life and Mind and Intelligence and Soul and Love and Fear and Joy and Sadness and Ideas and abstract Thinking and Creativity and Intuition, Imagination and Inspiration were all just fancy words for *atoms and chemicals!* The premise was faulty and it led to a perverse and false conclusion, but it's what they thought.

'But that doesn't make sense, Ollie. As you said, anyone who has ever seen a dead body of any kind can tell that something, that was there before the chemical body 'died', has departed, a pneuma, a prana, a chi, a shakti. It seems unreal to believe that Life is 'just chemicals' and that Intelligence and Minds and thoughts and Ideas and Feelings and Actions are 'made of' atoms. Does anyone really think that? And the idea that the existence of the Life Force was proved or disproved by the difference between the organic and inorganic chemicals was unfounded …'

'Go on …'

'It was wrong because to imagine that the existence of the Life Force could only be confirmed by the difference between organic and inorganic chemicals being a *fundamental* one, as opposed to a *conditioned* one, is as unreasonable as believing that the existence of TV signals and shows (or Life Forces and Souls) independent of a given TV (or the body) could only be confirmed by there being a *fundamental* difference between the metals and plastics in TVs (or the 'organic' chemicals) and the un-mined and unrefined chemicals in the ground, iron, copper, crude oil, etc (or the 'in-organic' chemicals) from which the metals and plastics in the TVs (or the bodies) are *intelligently* derived.

'No one believes that there can be no TV signals and TV shows apart from the metals and plastics in given TVs (our bodies) just because it turns out that the metals and plastics in TVs (the 'organic' or organ-ized chemicals) are not fundamentally different from the unrefined materials in the ground (the 'in-organic' or un-organ-ized chemicals) from which they are *intelligently* derived.

'It would be absurd to say that because people suddenly realized that the metals and plastics in TVs are fundamentally the same chemicals as the ores and oils that come out of the ground that this proved that TV signals and TV shows have no independent existence of their own and therefore 'must be' the 'chance' byproducts or 'acciden-tal' epiphenomena *of* and wholly dependent for their existence *on* the metals and plastics in a given TV which, Darwinian-style, 'accidentally' mined and refined themselves out of the ground and 'accidentally' self-assembled in 'accidental' TV making plants.'

'Yes, Alisha, you're right, it is unreasonable. When our TVs sometimes break down irreparably ('dead'), no one suggests that the TV Signals and TV Shows have also 'died', or that they never really existed in the first place. But this was the unreasonable conclusion drawn in 1828 after the artificial synthesis of urea, and it was declared that there were no Life or Vital Forces. This dreary conclusion also contributed to the *'abolition'* of the Soul, Life after death, and the 'super' Natural or Spiritual dimensions of Life as concepts within mainstream, western, scientific and even philosophical thinking which also became utterly materialistic in its approach. Our ancient western wisdom tradition went up in hallucinatory smoke.'

'Although, Ollie, the original theory of the Life Forces was not entirely unreasonable ...'

'Go on.'

'Look, the Vitalists were onto something, because of course there's a *difference* between the organic and in-organic chemicals, and the difference doesn't lie in their fundamental chemical composition anymore than the difference between the metals and plastics in TVs and the ores and oils in the ground lies in their fundamental composition. The difference lies in *what it is* that makes the organic chemicals (or the metals and plastics) different organ-izationally from the inorganic chemicals (or the ores and oils) – *what it is* that synthesizes the Urea.'

'Okay ...'

'It's *intelligent* forces, for heaven's sake! The organic chemical, urea, *didn't synthesize itself* out of inorganic chemicals in 1828. The metals and plastics in TVs don't mine and refine themselves out of the ground, by 'accident', Darwinian style. The inorganic chemicals of oxygen, carbon, hydrogen and nitrogen don't 'accidentally' turn themselves into daisies, buttercups, water melons, grapes, eagles and people. In the case of TVs *we* are the Life Forces, embodied in our *intelligent* intentions and desires, that turn the ores and oils in the ground (the *'in-organic'* or *un-organized* chemicals) into the metals and plastics (the *'organic'* or *organ-ized* chemicals) for use in TVs for the material-ization or transmission of the TV shows (our Souls in the metaphor).'

'OK Ali, good point. It is the in-form-ing and organ-izing *Life Forces* in and of the plants and microorganisms that turn the *inorganic* chemicals in the ground into *organic* chemicals for use by the plants and, later, the *Animal* and Human Soul-Beings. So, just as it takes *us* to turn 'inorganic' iron ore and other metallic ores and crude oil into 'organic' TVs or Rolls-Royces or Jumbo Jets it takes intelligent, higher Natural or Spiritual laws, Life and Soul Forces to turn unrefined and unelaborated inorganic materials into *anima*-ted (en-Souled) *Living, Breathing, Reproducing, Bio*-logical, Vital Force driven microorganisms, plants, animals and people by *vastly more sophisticated processes,* let it be said, than those that we use to turn the mineral ores and oils in the ground into TVs.

'The Victorian materialists, however, *did not agree* with this view and came to be of the firm opinion that vital, sacred, magical *Life* is *not really Life at all but is just dead, zombie chemicals having 'accidents'*.

'Of course, this doesn't work. This is one reason why we are in such a dead end today, philosophically and metaphysically, *because until biology recognizes* the Life and Soul Forces once more we'll continue not to have a clue as to *what Life! actually is,* and we'll persist with the impossible fiction that it's just some weird form of 'accidental chemistry'. This is not a reasonable approach to understanding the mystery of very Life itself. The reality is that nothing in biology or the Living World makes sense except in the light of the intelligently in-form-ing and organ-izing Life and Soul Forces.

'The abandonment of the logical inference of the Life Forces was, however, part of the intellectual *background* to Darwinian–materialist thinking. Firstly, the materialist idea that there are no Life Forces, therefore there probably is no immortal Soul or Life after death either. This was in complete disregard of collective humanity's nearly universal intuitions and witness that these are real. Didn't widespread and *real* human experiences and knowledge count for anything with those thinkers and their ideological descendants? Therefore, the now demoted pseudo 'Soul', pseudo 'Life' and pseudo 'Mind' were alleged to be merely 'accidental' epiphenomena of dead and allegedly 'purposeless' chemical matter.

'This led to the unreasonable and and by now many times refuted materialist idea that the Universal Witnessing Awareness, Subjectivity, Sentiency, Mind and the spiritual essences of Intelligence, Love, Purpose, Creativity, Goodness, Beauty and Truth are 'accidents' of chemicals and 'made' of them – chiefly atoms of oxygen, carbon, hydrogen, and nitrogen – as opposed to merely being transmitted by them into physical reality by incredibly *clever* bio-physical processes. Processes which make *all* the things we make, supercomputers included, and the methods we use to make them, look foolish in comparison.'

'But that is as rational, Oliver, as imagining that TV signals and TV shows are accidentally *'made of'* and accidentally *'made by'* the metals and plastics in TVs that 'accidentally' mined and refined themselves out of the ground, 'accidentally' transported themselves to TV making plants, 'accidentally' self-assembled there before 'accidentally' making the journeys into our homes, 'accidentally' setting themselves up for our use and 'accidentally' making TV Shows for us.'

'And the imaginary process you've just described, Ali, complex and impossible as it is, is far simpler than the 'accidental' and unintelligent-processes that would need to cause Living Cells ever to arise as a result of blind chance causes and eventually become today's Living World.[73] Yet, despite all these points, materialist thought has, at times, even argued that by looking among people's brain chemicals it will be possible to find immaterial ideas, which belong to the world of thought, Soul and Spirit, like Newton's Ideas for Differential Calculus or Beethoven's Ideas for one of his symphonies or Einstein's Ideas for the theory of General Relativity.

'This is as unreasonable as imagining that we could find the ideas and meanings of Shakespeare's plays and poems by analyzing the chemical composition of the paper and ink with which he wrote or by analyzing the chemical composition of the magnetic media on which his plays are stored on our computers, or that we could find TV Shows like the Sport or the News by analyzing the chemical composition of the wires, metals, circuits and plastics in the TV!

'The wires and circuit boards (the body) only transmit the Shows (the Mind and Soul) they don't create them. The paper and ink are only vectors for Shakespeare's ideas, just as deoxyribonucleic acid is but the ink or carrier for the *meanings,* intentions and *intelligent instructions* embedded in the digitized DNA CODING software that helps to make all physical-ized Soul Life Form Beings possible.'

'Yet, in materialist thought when we die, that's it?'

'Yes, of course. In the materialist scheme all we ever were was an unnecessarily complicated form of fertilizer. An outrageously complex but pointless form of Soulless, accidental, zombie chemistry. This is the unreasonableness that passes for an intelligent explanation of Life's origins and evolution in modern materialistic science and biology. A Soul corroding delusion or collusion that children and young people are exposed to all over the world, every day, and we wonder why they are so often disillusioned with Life – because a beautiful classical science captured and distorted by the ideology of materialism has told them, if never quite so directly and honestly, that:

> 'You, your joys and your sorrows, your memories and your ambitions, your sense of personal identity and free will, are in fact no more than the behavior of a vast assembly of nerve cells and their associated molecules. As Lewis Carroll's Alice might have phrased it: "You're nothing but a pack of neurons."'[74]

---

[73] Denton, M. 1985. *Evolution: A Theory in Crisis.* Burnett Books.
[74] Crick, F. 1995. *The Astonishing Hypothesis.* Scribner; p. 3.

6 Living Cells: Like Chemical Factories, Only More Complicated

How up to date Science, not religion, shows that the Living World is *not,* after all, an *'accident',* that this is not merely highly improbable but it's impossible.

'In 1871 Darwin famously imagined that Life might have had a chance beginning in "some warm little pond", "where a protein compound was chemically formed, ready to undergo still more complex changes" and become a tiny Living Cell, which Paul Davies describes as "the *most complex system* known" to humanity.

> 'Although the tiniest bacterial cells are incredibly small ... each is in effect a veritable *micro-miniaturized factory ... far more complicated than any machinery* built by man.'[75]

'Materialistic science notes the complexity but hypothesizes, nonetheless, that it is the product of 'accidental' forces. The scientific hypothesis of idealism, on the other hand, is that the laws of 'accidental' chemistry cannot possibly explain the numerous *irreducible* and *specific* complexities of intracellular and multicellular Life. Yet those who argue the idealist and intelligent design cases in biology today are met with tremendous resistance and even hostility. It is not as though they are saying:

> "We should just stop doing science and all go home because super Nature or Supreme Intelligent Beingness cleverly evolved the Living World and, just as important, sustains it, second by second, 24/7, week after week, for millions of years, so why bother studying anything."

'Would we lose interest in studying a previously unknown Rolls-Royce that appeared on our desert island, out of the blue one day, just because we realized, at some point, that it wasn't an 'accident' after all, and, that it had *intelligent and meaningful* origins and could not

---

[75] Denton, M. 1985. *Evolution a Theory in Crisis*. Burnet Books, p.250 emphasis added.

possibly have arisen as a result of random and accidental processes as some of the materialist thinking islanders had, at first, argued? All modern, non-materialist, intelligent design thinking proposes, in relation to evolution, is that unless we posit certain *higher level or 'super'* Natural laws, forces or processes, even if with current instrumentation we cannot physically detect them, we'll never be able to make full sense of the intelligent and purposeful, if challenging and mysterious, total Holistic Natural Reality in which we find ourselves.

'For example the key question of the Life or Vital Forces. If modern biology continues to deny that the Life forces or Life principles, that are known, in the East, by names such as Qi, Chi, Prana, Shakti and Kundalini, even exist, we'll never be able to make full sense of Bio-Logy, of the Logos of Life, of very 'Life' itself, because we'll continue to counterfactually insist that really there is no 'Life' there's just some kind of as yet ill understood form of accidental mineral chemistry.

'Yet methodological–materialism, in biology, *presupposes* the very thing it has not demonstrated. It dogmatically hypothesizes that random chemicals, floating around, could, let alone would, accidentally bump into each other, without any intelligent superordinate laws, forces or processes ever being involved, become initially 'simple' single Living Cells (which we now know are not 'simple' at all but amazingly complex, almost beyond belief), embodying Consciousness, Sentiency and Purpose, then, again 'by accident', become multicellular Life Forms, then become ever-more complex animal forms until, eventually, some of these animals *accidentally* turned into humans with the kind of (immaterial) *minds and intelligences* capable of working out differential calculus, writing symphonies, building space shuttles and suspension bridges, working out the theory of general relativity and using the amazing ontological and Soul quality we call *intelligence* to conclude that they *intelligently* emerge from a mysterious Cosmic matrix that is itself *utterly mindless and accidental.* If all we really were was Life-Forceless, Soulless, 'accidental zombie chemistry' we'd not be able to do these things. It should be obvious, Ali.

'Listen to this. William Dembski, a contemporary scientific–Idealist and leading intelligent design thinker has done some interesting mathematical research.'

'OK . . . go on.'

'He has found that even if it was possible to turn *the whole Universe* into a stupendously vast, chemical accident generator, a cosmically scaled Darwinian style 'warm pond' or 'primordial soup' so that every last particle, in the entire Universe, from the very beginning of time (14 billion years), was dedicated to doing nothing but randomly generating amino acid sequences, *it would **never**,* even under such ideal conditions, 'accidentally' *lead to just one of Life's thousands of intra-cellular protein molecules* forming in the necessarily correct sequence – *not even one.*'

'That sounds incredible, Ollie. Why?'

'This is because these *extraordinarily complex protein molecules,* vital to intra-cellular functioning, of which there are *thousands* in all cells, have to be in *precisely the right order* – (what is referred to as '***specified,*** intelligent complexity', not 'random, meaningless complexity', like rocky, rubbly screed or jumbled up scrabble letters) – or they will not work.

'However, Dembski has found that the random, chance generating resources of the *entire universe* are not large enough to 'accidentally' get even *just one* of these proteins assembled in the right order, even in a universe-sized pond, there from the beginning of time, 14 billion years ago. In any case, Ali, a Cell is many orders of magnitude more complex than *just one* of these proteins, and even if one ever did arise by chance – be it in Darwin's "warm little pond," or even in a universe sized pond, *with no cellular membrane* and other vital features of *self-directing* cellular Life to protect it, it would be as rapidly destroyed by the other *accidental-chemical-reactions* going on around it 24/7.

'Dembski says, however, that a mechanism's Specific Complexity should be high before the materialist hypothesis of random, chemical chance can be completely ruled out as a possible explanation of its origin. He explains that, at some point, a *specific* and *meaningful sequence* of information, as we see in Life's proteins, made of specifically ordered and folded polypeptide chains, "becomes too long, too improbable, *too complex* to reasonably attribute to *unspecified,* unintelligent chance" [76]

'In their book *Intelligent Design, Uncensored,* Dembski and his co-author Witt, ask:

> 'How improbable does a specified thing have to be before we can know it was designed? One chance in ten is obviously too low a threshold. One chance in a hundred is too low. How high is high enough? For our purposes, we don't mind if the test occasionally misses design. We just want to make certain the test doesn't say something was designed that wasn't. This means we want to set the bar very high, meaning the thing in question will have to be ***extremely improbable*** to pass our design test.' [77]

---

[76] Dembski, W and J. Witt. 2010. *Intelligent Design Uncensored.* IVP, p. 66.
[77] Ibid. p.67. Emphasis added.

They say, '*The Design Inference*[78] sets the threshold at 1 chance in 10 to the power of 150.'[79] That is incredibly high Ali – it means 1 chance in 1,000,000,000,000,000,000,000,000,000,000,000,000,000,000,000,000,000,000,000,000,000,000,000,000,000,000,000,000,000,000,000,000,000,000,000,000,000,000,000,000,000,000,000,000,000,000,000,000,000. They continue, "If a monkey can win a prize by randomly typing the name George, he doesn't have much of a chance if he only has a few tries, since the odds on any given try are less than one chance in 300 million." But, if he did this for "nine hours a day for a year, he has a small but realistic chance of hitting the name".[80] In "the known physical universe there are about 10 to the 80 [1 + 80 zeroes] elementary particles". Planck Time shows that matter can change from one state to another no faster than 10 to the 45 per second [1 + 45 zeroes].

> 'If we now assume that any event in the universe requires the transition of at least one elementary particle … then the total number of [such transitions or] events throughout cosmic history [14 billion years] could not have exceeded 10 to 80 x 10 to 45 x 10 to 25 = 10 to 150.
>
> This means that any specified event whose probability is less than 1 chance in 10 to 150 will remain improbable even if we let every corner and every moment of the universe roll the proverbial dice. The universe isn't big enough, fast enough or old enough to roll the dice enough times to have a realistic chance of randomly generating specified events that are this improbable.[81]

'We should note that Dembski's probability test is the most demanding in the literature. Other researchers put the universal probability threshold much lower, Cryptographers at 10 to 94, French mathematician Emile Bore at 1 chance in 10 to the power of 50. Dembski and Witt say that *all* of these figures are, "far, far higher than what would normally be required to safely infer design".[82]

'They continue:

> '*The Design Inference* (1998) set the bar even higher to be doubly sure that we avoid labeling anything as *designed* that wasn't

---
[78] W. Dembski, *The Design Inference: Eliminating Chance through Small Probabilities.* Cambridge Studies in Probability, Induction and Decision Theory. Cambridge University Press, 2006.
[79] Ibid., p. 67.
[80] Ibid., p. 68.
[81] Ibid., p. 69.
[82] Ibid., p. 69.

designed... *we now know that even the simplest bacterial cells are **complex** almost **beyond belief**.*[83]

'And, quoting Michael Denton:

'Although the tiniest bacterial cells are incredibly small... each is in effect a veritable micro-miniaturized factory... made up altogether of one hundred thousand million atoms, *far more complicated* than any machinery built by man ...

'They say this makes it clear that even:

'... the simplest cell is **specified**, that is, it conforms to an independent pattern. In this case the pattern is that of a self-reproducing factory for producing functioning hardware and software ... [It is] irrational to behold such a sophisticated factory and say, "So what? It's just a bunch of chemicals".[84]

'Yet, even so, Dembski and Witt ask: 'Is the arrangement of the parts of the most 'basic' Cell, complex enough to pass his rigorous mathematical test for detecting intelligent and purposeful functioning in Nature versus 'accidental' design?' They say:

'Scientists calculate that a cell with just enough parts to function in even a crude way would contain at least 250 genes and their corresponding proteins. The odds of the early Earth's chemical soup randomly burping up such a micro-miniaturized factory are *unimaginably longer* than 1 chance in 10 to 150.'[85]

'So even with trillions more chances than would ever be available in some ancient Darwinian sea or pond, there is no way Life on Earth could ever have arisen by unintelligent or accidental processes?'

'Yes, exactly, Ali. That's the point they are making. Dembski and Witt explain that even if just one of Cellular Life's thousands of proteins did somehow accidentally arise in a hypothetical primordial soup it would be immediately destroyed since the random protein string would "lack the cell's **self-directing** and protective structure."

---

[83] Ibid., p. 69.
[84] Ibid., p. 71.
[85] Ibid., p. 71.'

'Antony Latham says:

> 'Far too often I hear experts confidently postulate about the first cell membranes. They know that the first life had to have a protective coat to hold the DNA and all the incredibly sophisticated molecular machinery inside the bacteria. ... [yet] the actual details of the membrane must be coded for in the DNA of the cell.'[86]

'As Latham points out, how could the first DNA CODING arise in some purely unguided and random way without first having the protective cellular membrane around it?'

'So it's chicken and egg, which came first?'

'Exactly, and these chicken and egg problems befuddle all the materialist attempts to imagine irreducibly complex Life arising 'by accident'. Of course, if Life is an intelligent, 'higher' Natural phenomenon then we don't have to waste time trying to work out how it arose 'by accident' and can, instead, try to work out how it arose, and as importantly, how it arises right now, second by second, 24/7, by stunningly *intelligent* processes. Currently, materialistic science and biology don't actually know what 'Life' truly is and content themselves with the materialistic belief that it's just 'accidental' chemistry regardless of the fact that that is not merely improbable, it is statistically and mathematically impossible.

'Some Darwinians argue that the information rich molecules that make Life possible might arise by accident where a non-living mixture of relatively simple proteins and RNA molecules might reach a point at which the mixture could self-replicate. This is called 'RNA World' origin of Life thinking. But why would such a mixture start to replicate? What for? Without a directive Life-Force of some kind what would cause such replication? Things only replicate if it is in their programed nature to replicate. This, however, begs the question of the source of such intelligent programming and it doesn't tell us what Life actually is. Are the Life Forms Soul Beings? Or are they 'accidental' mineral chemistry 'by other means'? Really just a branch of physics? Accidental, pseudo–creature, selfish-gene machines that exist to replicate to replicate for no reason at all other than to replicate! This is what materialism does to our scientific culture, Ali, it makes it irrational.

'Although Darwin was unable to produce even "*one* bona fide case of natural selection having *actually* generated evolutionary change in nature ..."[87] we are asked, in the imaginary 'RNA world' origin of Life

---

[86] Latham, A. *The Naked Emperor: Darwinism Exposed*. London: Janus, p. 13.
[87] Denton, M. 1985. *Evolution: A Theory in Crisis*. Burnett Books, p. 62. Emphasis added.

scenario to believe that 'natural selection' would act on such an imaginary mixture to turn it into a Living Cell. This is as reasonable as arguing that a pile of TV or Car parts might 'accidentally' self-assemble. In any case RNA molecules *already contain complex specified information* the origin of which remains unexplained in the materialist 'RNA world' scenario. As William Dembshi explains, in *No Free Lunch: Why Specified Co-plexify Canot Be Purchased Without Intelligence* (2007), there is no 'free' instructional information – a problem that no purely materialistic theories of Life's origin can overcome. Yet all Life Forms are run by sophisticated digitally CODED information and information systems that make our supercomputers and their software codes look like wooden spoons in comparison.

'In fact, Dembski and Witt say that the chances of the notional primordial soup generating a *self-directing* and *protective* structure for the cell at the same time and in just the right place as any random protein string, (which would be just one of the thousands required to make intra cellular life possible), is *beyond improbable*. They say that it is so high that:

> 'it stretches the ability of biochemists to calculate it. The best they can do, is set a lowest possible figure, which *surpasses by untold trillions of trillions of trillions of times the universal probability bound of 1 chance in 10 to 150.*'[88]

'The reality is, Ali, that specifically and irreducibly complex, (not randomly complex) Living Cells, which make all the larger Life Forms possible, and here from the very beginning, 3.8 billion years ago, all perfect and complete, with no signs at all of any Darwinian–style, gradually mounting rise to increasing complexity, could *never, ever,* have arisen by pond-water *accidental-chemical-reactions*. The materialist idea that Life is a purely 'material' and an 'accidental' phenomenon *does not work*. It is not merely highly improbable, as everyone has always agreed – it has now been shown to be ***impossible***.

'So why doesn't the mainstream scienfitic community take this on board?'

'Because of, as I keep saying, its commitment to exclusively materialist or ('ordinary') naturalist (that is non 'super' naturalist) explanations of all phenomena. This is what methodological materialism, sometimes also referred to as methodological naturalism, is. The knowledge blocking and science distorting insistence that only the known laws

---

[88] Dembski, W and J. Witt. 2010. *Intelligent Design Uncensored*. IVP, p. 72.

physical nature, the laws of physical chemistry, are real and they are fully sufficient, they *have to be* fully sufficient, regardless of logic, rationality or all the evidence to the contrary, to explain all phenomena.'

"Yet, as Michael Behe also demonstrated in his world-class book *Darwin's Black Box: The Biochemical Challenge to Evolution* this does not work. Behe pointed out that no one has ever explained in a serious or plausible scientific way how random genetic mutations and natural selection could build the stunningly complex biochemical and molecular structures he discusses in his book, such as the irreducibly complex blood clotting cascade, the irreducibly complex bacterial flagellum and the irreducibly complex process of protein synthesis. He says:

> 'In fact, evolutionary explanations even of systems that do not appear to be irreducibly complex, such as specific metabolic pathways, are missing from the literature. The reason for this appears to be similar to the reasons for the failure to explain the origin of life: a choking complexity strangles all such attempts.'[89]

'However, we still live in 'pre-Galilean' times and the majority view in biology remains materialistic, that is that Life is a 'no–intelligence–involved–accident'.

Ali looked baffled, she said,

'I still don't understand, Ollie, why science and biology are in such unreasonable intellectual thrall to the materialist perspective which violates the classical *logic* that intelligent mechanisms only ever have intelligent sources, never 'accidental' ones. Why is it also so reflexively dismissive of collective humanity's nearly universal intuitions, psychic experiences and mystical and spiritual visions and revelations that, ultimately, Life and the Living World and our existences actually mean something and are sourced in dazzling, scintillating, blazing *intelligence* and *purpose* – not mindlessness, unintelligence and 'accident'?'

'I think, Ali, that a lot of scientists imagine that if we conclude that Nature is *intelligent* and *purposeful* we're not being 'scientific' anymore. We've given up on finding so-called purely 'material' or 'ordinary' Natural explanations for the Living World. We've succumbed to unwarranted 'super' Naturalism or multi–dimensionalism. But this methodologically–materialist view *assumes and presupposes the very thing it needs to prove*, which is that Nature and Existence are *not*

---

[89] Behe, M. J. 1996. *Darwin's Black Box: The Biochemical Challenge to Evolution*. Free Press; p. 177.

intelligent, are not *intelligently* multi-dimensional or, ultimately, 'super' Natural phenomena. Assuming they are not – for example because we cannot physically see but can only rationally infer or intuit Life's 'super' or higher Natural Final Sources and Causes – is not proof of this foundational materialist contention; it's just a metaphysical prejudice or belief. Sadly though, it's the dogma that governs biology today.

'On the other hand, it was Holistic Nature's very rationality and *intelligence* that caused the founding giants of western science and philosophy – Plato, Aristotle, Bacon, Galileo, Newton, Faraday, etc – to reason that it might be possible to do science in order to explore this very *intelligence* and to come to intelligent grips with it. These giants of thought saw Nature as *intelligent* and *purposeful, not random and meaningless*. They saw it as containing higher or 'super' Natural elements and dimensions as well as the so-called 'ordinary' Natural elements, that we take so for granted, and, therefore, the concept of intelligent design in Nature was a logical corollary of that view. This has also been the nearly universal intuition of collective humanity, not to mention the spiritual revelations of countless people, some of them even epoch-making – Zoroaster, Moses, Lao Tzu, Buddha, Jesus, Mohammed, Shankara, Guru Nanak, and so on.

'Sir Frederick Hoyle prefigured Dembski when he said:

> 'Life **cannot** have had a random beginning … the odds could not be faced even if the whole universe consisted of organic soup.'

'He rightly realized that the reasons for resisting the reasonable conclusion that we live within a multi-dimensional Cosmos of Intelligence and Meaning are "psychological rather than scientific". Dembski's mathematical work, which he details in *The Design Inference* and Behe's in *Darwin's Black Box* both show, just as Hoyle concluded, that our mysterious and magical Living World of the amazing Plants, Animals and People simply *cannot be* the result of a billions year long running *'accidental chemical reaction'* – it's not merely incredibly improbable, the old Darwinian or 'pre-Galilean' view in biology, but *impossible*. It is clear, today, that the strange materialist love affair with the impossible twin myths of an 'accidental' Living World and a mindless Cosmos of unintelligence and unmeaning is unfounded.

We continued on with our pleasant morning walk through the enchanting valley with the refreshing river by our side and the beautiful fields and meadows all around.

7 Evolution: Intelligent Unfoldment or Random Fluke?

> Darwin's idea that Life and the amazing Living World are evolutionary 'accidents' was unrealistic – impossible in fact. Living Nature's 24/7–dynamic bio-functionalities only start to make sense when we drop the idea that they are 'chance' processes and acknowledge that the Fossils make visible a wonderfully *intelligent* unfoldment of Life across vast time.

'Ali, if we said "Just Chemicals + Accidents = Wooden Toy" we'd be absurd. If we said "Just Chemicals + Random Events = Jumbo Jets" we'd be bonkers. Yet, today, if we say "Just Chemicals + Accidents = tiny Living Cells, Plants, Animcals and People, all vastly more complex than Jumbo Jets" we are supposedly being 'scientific' and there is much 'Naked Emperor' type clapping and applause.'

'So why do so many people accept the materialist view of Living Nature, as a Life Forceless and soulless *'accidental chemical reaction'* that's been running for no rhyme or reason for millions of years?'

'Partly, I think, because it is what we are all taught. Yet as Jonathan Wells explains in *Icons of Evolution, Science or Myth? Why Much of What We Teach About Evolution is Wrong*, there are a lot of myths and even outright deceptions in the promotion of Darwin's theory of blind, chance evolution. For example, Haeckel's famous but faked embryo drawings - still, to this day, reproduced in some textbooks. The famous Miller–Urey experiment deceptively promoted as evidence that Life might have started 'by accident' when all it produced was some of Life's most basic building blocks. This was like heating and tumbling a barrel of clay and water and claiming that the resultant fired clay pebbles prove that ancient Rome, built out of some of the same basic materials also arose by a similarly random and unintelligent process. The misleading claim that minor or so-called micro-evolutionary finch beak and peppered moth variations (which involve no genetic mutations at all, let alone random ones) are good evidence for Darwin's hypothesis of macro-evolution – that astonishingly complex microbial Life could, in the first place, arise by chance and then – blindly, randomly and unintelligently – some such hypothetical microbe could accidentally turn into a trilobite or fish and then an amphibian, dinosaur, bird, mammal or, finally, human beings with intellectual abilities vastly in excess of the mere need to survive and replicate genes.[90]

'It is also the case that, senior, world-famous scientists regularly put out books reinforcing the materialist opinion that Living Nature is simply a very long running 'accident'. So what are we to think? The great and the good of the mainly materialist thinking scientific world tell us that Life is a random fluke and that there are no Life Forces or Souls or Life after death because these are not physically detectable but can only be *rationally inferred* or detected by non–physicalist methodologies and hypothesis testing. To quell our child-like 'But the Emperor is Naked' doubts, they insist that Darwinian Science has conclusively shown this. The case is settled and closed.

'This, however, is not actually true. It's a bluff. A motive could be professional scientific pride, not wanting to admit that the strange, gloomy, materialist thought project, started by Darwin and his Victorian materialist contemporaries, has run into the sands and is just 'wrong'. Look, even Darwin, himself, said,

> '... The number of intermediate varieties, which [if his theory was correct, must] have formerly existed on the earth, (must) be truly enormous. *Why then* is not every geological formation and every stratum full of such intermediate links? Geology assuredly does not reveal any such finely graduated organic chain; and this, perhaps, is the *most obvious and gravest objection which can be urged against my theory*'.[91]

He also said,

> 'I am well aware that there is scarcely a single point discussed in this volume on which facts cannot be adduced, often apparently leading to conclusions *directly opposite* to those at which I have arrived. A *fair result* could be obtained only by *fully stating* and *balancing* the facts on *both sides of each question,* and this cannot possibly be done here.'[92]

'Yet, when, today, scientists and philosophers put forward credible, idealist theories, such as those of intelligent design, "leading to *conclusions directly opposite* to those at which [Darwin had] arrived" they are resisted and mocked by those who insist that ongoing, Living Nature of stunning dynamism, trillions of astonishingly complex

---

[90] Wells, J. 2002. *Icons of Evolution*. Regnery Publishing.
[91] Darwin, C. 1859. *On the Origin of Species*. Emphasis added.
[92] Ibid.

operations occurring in our bodies this very instant, every instant, is a second by second, 24/7 'accident' that only 'looks' blazingly clever – 'actually it's not. Really it's without intelligence, purpose or meaning.' Art Battson says:

> 'Charles Darwin was well aware that scientists could come to *directly opposite* conclusions from those set forth in his *Origin of Species*. Although his theory could account for *minor* evolutionary change and the diversity of finches, Darwin knew that he had to *virtually ignore* the natural history of life on earth in order to maintain any hope of accounting for the origin of the phyla [basic anatomical designs or body plans] and the major disparity between arthropods and anthropologists.'

'He's right because Darwin's theory totally conflicts with the most basic aspects of natural history. Firstly the Fossil Records *do not provide any of the transitional forms* that his theory predicted. In fact quite the reverse is the case, the species typically arise suddenly and fully formed, most dramatically in the case of the Cambrian explosion, with no evidence for the gradualistic emergence of new anatomical designs that Darwin had in mind. 'In his book, *The Naked Emperor: Darwinism exposed* Antony Latham comments on world famous paleontologist, the late S. J. Gould's discussion concerning Darwin's problem of the sudden, discontinuous appearance of multicellular animal life in the Cambrian. He quotes Gould:[93]

> 'Darwin invoked his standard argument to resolve this uncomfortable problem: the fossil record is so [allegedly] imperfect that we do not have the evidence for most of life's history. But even Darwin acknowledged that his *favorite ploy was wearing a bit thin* in this case. His argument could easily account for a missing stage in a single lineage, but could the agencies of imperfection *really obliterate **absolutely all evidence*** for positively every creature during most of life's history? Darwin admitted: *the case at present must remain inexplicable; and may be truly urged as a valid argument against my views here entertained (Origin of Species, 1859).*

'Darwin has been [according to Gould] vindicated by a rich Precambrian record, all discovered in the last 30 years. Yet the peculiar character of this evidence has not matched Darwin's prediction of a continuous [gradualistic, accidental] rise in

---

[93] Latham, A. 2005. *The Naked Emperor: Darwinism Exposed.* London: Janus, p. 39.

complexity toward Cambrian life, and the problem of the Cambrian explosion has remained as stubborn as ever – if not more so, since our confusion now rests on knowledge, rather than [Darwin's] ignorance, about the nature of Precambrian life.'[94]

'Antony Latham reflects on this. He says:

'Gould tries to have it both ways, saying Darwin has been vindicated but then *refuting this* by pointing out the fossil record that we *do* have shows ***no Darwinian gradual progression*** of animals. As already described, we do know of fossils before the explosion but these ***completely overturn the Darwinian model***. For 3.2 billion years [that is from 3.8 billion years ago to 600 million years ago] we have only microbial life [with no Darwinian progression of animals "accidentally" evolving from unicellular bacterial complexity to ever greater multicellular complexity]... Then, in a period of about 50 million years: the Ediacaran Fauna which appear unrelated to the Cambrian fauna followed by the host of body plans of the Cambrian explosion. Let me quote Gould again from later in the same chapter of his book:

"Thus, instead of Darwin's gradual rise to mounting complexity, the 100 million years from Ediacara to Burgess may have witnessed 3 radically different faunas – the large pancake-flat soft-bodied Ediacaran creatures, the tiny cups and caps of the Tommotion, and finally the modern fauna, culminating in the maximal anatomical range of the Burgess. Nearly 2.5 billion years of prokaryotic [bacterial] cells and nothing else – two-thirds of life's history in stasis at the lowest level of recorded complexity [which is, as Denton explained, spectacularly complex and not "simple" at all]. Another 700 million years of the larger and much more intricate [even more complex] eukaryotic cells, but *no aggregation to multicellular animal life* [which, again, Darwin's theory predicted]. Then in the 100 million year wink of a geological eye, three outstandingly different faunas – from Ediacaran, to Tommotion, to Burgess. Since then, more than 500 million years of wonderful stories, triumphs and tragedies, but ***not a single new phylum, or basic anatomical design***, [which, once more, Darwin firmly predicted should be the case if his theories were correct], added to the Burgess complement.

---

[94] Gould, S. J. *Wonderful Life. 1990. The Burgess Shale and the Nature of History*, New York: Vintage, p. 57. Emphasis and text in square brackets added.

Step way back, blur the details, and you may *want* to read this sequence as a tale of predictable progress: prokaryotes [non nucleated bacterial cells] first, then eukaryotes [nucleated single cells], then multicellular life. But scrutinize the particulars and **the comfortable [Darwinian] story collapses. Why** did life remain at stage 1 for *two thirds* of its history if [accidental] complexity offers such benefits? **Why** did the origin of multicellular life proceed as a short pulse through three radically different faunas, rather than as a slow and continuous rise of complexity [as Darwin predicted would be the case if his theory was correct]?"[95]

'So what are the implications of this, Ollie?

'The implications are, once more, that purely materialist or non 'super' Naturalist, no-intelligence-involved theories of Life on Earth's origins and its subsequent evolution or unfoldment through time cannot explain the data. Anti-super Naturalist Darwinism needs smooth continuity but the Fossils do not oblige. The species come and go from the records discontinuously or 'jumpingly', resembling something more like new car model evolution where each new 'species' of car suddenly appears (as a result of intelligent causes), stably lasts for a while and disappears.

'When Living Nature is understood, once more, to be a 'super' or 'higher' Natural phenomenon involving *intelligent* input the discontinuities or 'jumping' nature of the arising of the Species make sense. There is then no need to continue to try to make the fossil records fit Darwin's theory when they so clearly do not and never have. The Fossil Life Forms show none of the Darwinian changeability and transmutability after their sudden, 'new car model' like appearances, but a *profound stability* or equilibrium during their periods of existence on Earth. There is no evidence for their gradually transforming, Darwinian style, into new creatures during their periods of existence.

'In *Darwin's Enigma* Luther Sunderland quotes Colin Patterson, a senior paleontologist at the British Natural History Museum, with whom he was discussing the lack of transitional species in the fossil records showing new anatomical features gradually developing, rather than appearing 'suddenly' and fully formed, a fact which, as S. J. Gould put it, is the "trade secret of paleontology." When Patterson was asked by Sunderland why he did not show any such transitional forms in his, Patterson's, book he said:

---

[95] Gould, S. J. *Wonderful Life. 1990. The Burgess Shale and the Nature of History*, New York: Vintage, pp. 59–66. Emphasis and text in square brackets added.

Evolution: Intelligent Unfoldment or Random Fluke?

"I fully agree with your comments on the lack of evolutionary transitions in my book. If I knew of *any [such Darwinian-style transitions]*, fossil or living, I would certainly have included them. I will lay it on the line – *there is not one such fossil* for which one might make a watertight argument."[96]

'As Richard Milton, a scientific journalist and non-creationist skeptic of Darwinian evolutionism, writes:

'no transitional species showing evolution in progress between two stages has ever been found. ... the theory of evolution has become *an act of faith* rather than a functioning science. ... Not until the scientific method is applied to Darwinism will it be exposed, and only then will the right questions be asked about the *mystery of life on earth.*[97]

'Battson points out that the actual nature of the Fossil succession:

'suggests the existence of natural processes which [actively] *prevent* major evolutionary change from occurring on a gradual step-by-step basis.' [98]

'Far from confirming Darwin the fossil records refute his theories. Astonishingly, given the unquestioning acceptance of Darwinism by the mainstream scientific community, the fossil succession is, in fact, *systematically backwards* from that predicted by his theory. Darwin predicted that the gradual accretion of minor evolutionary changes and the increasing variations in the earlier Life Form groups or 'taxa' should, eventually, lead to the great differences among the basic anatomical designs or 'phyla' of the Life Forms. Minor variations, Darwin predicted, should *precede* disparity, that is the massive differences in basic body plans. Yet the Fossil succession reveals precisely the opposite: massive differences in basic body plans arrive all at once in the Cambrian entirely preceding the subsequent more minor variations within the basic body plans.

---

[96] Sunderland, L. 1988. *Darwin's Enigma: Ebbing the Tide of Naturalism.* p. 89. Emphasis added.
[97] Milton, R. 2000. *Shattering the Myths of Darwinism.* Inner Traditions Bear and Company. Emphasis added.
[98] Battson, A. *On the Origin of Stasis by Means of Natural Processes.* Battson's website. Emphasis and text in square brackets added.

'This means, Ali, that Darwin's tree of life turned out to be a 'hedge', not one tree, but many trees, representing the basic anatomical designs, which arrived all at once in the Cambrian, and that far from the original 'trees' or body plans adding yet more new versions over time, as Darwin's theory predicted, many of them died out. The hedge of the Living World's basic anatomical designs or phyla is *smaller* today than it was in the Cambrian, the precise opposite of what Darwin's theory claimed. Battson says:

> 'Had Darwin developed a theory to explain the empirical data of natural history, he should have come to *directly opposite conclusions*. He should have developed a theory to explain *why species do not gradually transform* into substantially different body plans [or basic anatomical designs] on a gradual step-by-step basis. The phenomenon of stasis and the ***stability*** of the major body plans is based upon an abundance of [real-world] data and our theories describing the natural world *should explain* that [real world] data.' [not seek to explain it away.][99]

'Battson says, in another article:

> '... if ***stasis*** [not continuous change as Darwin predicted] is the principle feature of individual species in the natural history of life, if natural selection [actually] inhibits or *precludes* major evolutionary change, if the order of appearance in the geologic record is ***systematically backwards*** to Darwinian predictions, and if the higher taxa are as *discontinuous* as they appear to be, *why is it* that scientists don't develop theories to explain the natural ***limits*** to biological change? The answer seems to be more philosophical than empirical. Although the question would probably lead to a more accurate description of nature, it *would undoubtedly undermine* the pervasive secular *philosophy of ['accidental'] mechanism and materialism* which has come to dominate modern science.'

'Battson notes that Michael Ruse, a philosopher of science and a noted defender of Darwinism, once said that modern evolutionary theory is almost like a materialist religion:

> "And... evolution has functioned as something with elements which are, let us say, akin to being *a secular religion* ... And it

---

[99] Battson, A. *On the Origin of Stasis by Means of Natural Processes*. Battson's website. Emphasis and text in square brackets added.

seems to me very clear that at some very basic level, evolution as a scientific theory makes a commitment to a kind of ['ordinary'] naturalism, namely, that at some level one is *going to exclude miracles* [or 'super' Naturalism] and these sorts of things *come what may*." (Ruse 1993)[100]

'The trouble is, Ali, this posture is shockingly unscientific.'
'Why?'
'Firstly, because science should not be an ideology or, as Ruse puts it, a *secular religion!* The problem is that for this kind of 'scientific' thinking "materialism is not a scientific hypothesis *that is open to potentially being proved false; it is an article of faith that "must" be true,* regardless of evidence to the contrary."[101] If this were not the case do you think there would have been so much mockery of and resistance to the modern idealist theories of intelligent design in mainstream science and biology?

'For example Michael Behe's books, *Darwin's Black Box* and *The Edge of Evolution* and Stephen Meyer's *Signature in the Cell: DNA and the Evidence for Intelligent Design* are impeccable in their heavy duty scientific credentials. Their scientific proposals, however, cannot be accepted by an ideologically materialist science that is, as Lewontin puts it in a 'struggle' with the 'super' Natural or intelligently causal, an ideologically materialist science that has:

> 'a prior commitment, a commitment to **materialism.** ... [an] **a priori** adherence to material causes [alone and] *material* explanations [alone]. ... [A science where] that **materialism is absolute**, for we cannot allow a Divine foot, [or *intelligent* causation], in the door.'[102]

'By contrast with Ruse and Lewontin, Chris Carter reminds us that:

> 'the difference between **science and ideology** is not that they are based on different dogmas; rather, it is that scientific beliefs are **not** held as dogmas, but **are open to testing and hence possible**

---

[100] Battson, A. 1997. *Facts, Fossils, and Philosophy.* Author's website. Emphasis and text in square brackets added.
[101] Carter, C. 2010. *Science and the Near Death Experience.* Inner Traditions; p.237. Emphasis added.
[102] Richard Lewontin (1997) Billions and billions of demons (review of *The Demon-Haunted World: Science as a Candle in the Dark* by Carl Sagan, 1997). *The New York Review*, January 9, p. 31. Emphasis and text in square brackets added.

rejection. Science cannot be an objective process of discovery if it is wedded to *a metaphysical belief* that is accepted without question and that leads to the exclusion of certain lines of evidence [such as the modern evidence for intelligent design or intelligent evolution] on the grounds that these lines of evidence contradict the metaphysical belief.'[103]

'As Bruce Greyson says:

"It is the scientific method of empirical hypothesis testing, rather than a materialistic philosophy, that has been responsible for the success of science in explaining the world. If it comes to a choice between empirical method and the materialistic worldview, the true scientist will choose the former"[104]

'Secondly, Ruse's statement that science "is going to exclude miracles [or intelligent causation in Nature] ... *come what may*" doesn't make sense because the fact that anything at all exists rather than nothing at all is *the greatest miracle or discontinuity of all!* We can call it 'dumb chance' if we wish but neither description, 'miracle' or 'accident' takes away from the 'impossibility' that there should be anything at all rather than nothing at all. Why should there be even be one quark? Where could it come from? Who or what could make it? Who or what could make the who or what that could make the quark or a Big Bang giving rise to a Universe that is 20 billion light years across? What kind of blazing intelligence or mindless, unintelligent, accidental-universe-generating-mechansm (MUA–UGM) could give rise to all that is? All we can decide is whether this impossible thing, this 'miracle' that existence is, is a 'mindless' miracle or a 'brilliant' miracle. Ideological materialism opts for mindless chance, philosophical idealism for blazing intelligence.

'The problem we face, however, in our attempts to return to a Cosmos of intelligence and meaning in our understandings of Nature is that the *whole point* of the Victorian–materialist thought–project was, and remains, to try to explain the changes, in the Living World, over vast time, that the Fossils make visible, as purposeless, 'intelligence-not-needed' processes – perhaps due to a gross over-reaction to

---

[103] Carter, C. 2010. *Science and the Near Death Experience.* Inner Traditions; p.237. Emphasis added.
[104] Greyson, B.: "Commentary on 'Psychophysiological and Cultural correlates Undermining a Survivalist Interpretation of Near-Death Experiences.'" *Journal of Near Death Studies* 26, no. 2 (Winter 2007); p. 142.

medieval 'god-is-a-man-with-a-beard' biblical literalism – a very unsophisticated concept of the *higher* Natural or intelligently causal: subtle, inter-dimensional worlds of teeming spiritual activity, countless subtle energies, rays, laws, forces, processes, purposes and agencies, synergistically nested hierarchies of intelligent, holistic, 24/7 functioning at numerous vibrations, dimensions and frequencies of which the outer, 3rd Dimensional, physical world is but one.[105]

'The good thing is that the rest of us who are *not* committed materialist theorists do not need to discard Darwin's useful insight that Life definitely *evolves*, even if we *do* drop his materialistic assumption that it does so by purely *unintelligent* and 'accidental' processes. As many people have pointed out, Darwinism is really much more an expression of *materialist philosophy* than it is a good explanation for Life's unfoldment.

'The only examples of 'natural selection' that have ever been found are extremely minor ones like variations in finch beaks or peppered moth populations which involve no genetic mutations at all, and are reversible, and very simple mutations in bacteria which are also reversible. These phenomena are known as 'micro-evolution' and no one disputes them. There is no evidence, though, for 'natural selection' being able to turn an allegedly 'accidental' bacterium into larger Life Forms, which is known as 'macro' evolution or large scale evolution. As Battson says:

> '... Darwinists have been less concerned with the *scientific* question of accurately explaining the empirical data of natural history and more concerned with the religious or *philosophical question* of explaining the design found in nature without a designer. Darwin's general theory of evolution may, in the final analysis, be little more than an unwarranted extrapolation from microevolution *based more upon [the] philosophy [of materialism] than fact*. The problem is that Darwinism *continues to distort* natural science.[106]

'The Fossil Records, in complete conflict with Darwin's predictions, show the discontinuous, 'new car model' like, jumping or punctuated unfoldment of the Living World over millions of years. What we *do not*

---

[105] Tompkins, P. 1997. *The Secret Life of Nature*. Harper Collins. See also the works of theosophy and R. Steiner's anthroposophy.
[106] Battson, A. 1997. *Facts, Fossils, and Philosophy*. Author's website. Emphasis and text in square brackets added.

yet know, despite Darwinian and materialist claims to the contrary, Alisha, is precisely *what drove* those changes. As Antony Latham says in his book, *The Naked Emperor: Darwinism Exposed*:

> 'What is *not in any way explained*, [by Darwinian theory] however, are the big changes that we call macro evolution these are the sudden appearance of completely new forms and structures [bacteria to fish, bacteria to apple tree, dinosaur to bird etc] – indeed, all the appearances of the phyla [basic anatomical designs] and classes (higher taxa). *These remain a total mystery and do not fit in with any known mechanism or theory.*[107]

'Darwinian theorists of 'evolution by accident' and unintelligent design continue to hypothesizes that the changes over time that the fossils reveal, were driven by purely *random* processes, by chemical 'accidents', random genetic variations. They are unwilling make any *rational inferences* to *intelligent causation* in and of the Living World because of their passionately held materialist and *anti 'super' naturalist* convictions. Battson says: 'Writing in the introduction to the 1956 reissue of the Origin of Species, W.R. Thompson commented:

> 'The success of Darwinism was accompanied by *a decline in scientific integrity*. This is already evident in the reckless statements of Haeckel and in the shifty, devious and histrionic argumentation of T. H. Huxley … To establish the *continuity* required by the theory, historical arguments are invoked even though historical *evidence is lacking*. Thus are engendered those fragile towers of hypotheses based on hypotheses, where fact and fiction intermingle in an inextricable confusion.'

'You see, Ali, there is *no way* that a classical, metaphysically neutral, non-ideological science would have *a priori* ruled out intelligent causation in or of Living Nature, as Darwinian and materialist biology both do, when *all* our experience tells us that unintelligent processes and 'accidents' are only good for explaining *unintelligent* and accidental effects like car crashes, rust, frost, random forest fires or randomly evolved landscapes and pretty river valleys – and *never* clever mechanisms or machines of any kind, let alone the sophisticated software codes needed to run them.

We walked on quietly for a while, then Alisha said,

'One thing, though, Ollie, won't many people say in relation to your

---

[107] Latham, A. *The Naked Emperor: Darwinism Exposed*. London: Janus, p. 59. Emphasis and words in square brackets added.

arguments: 'Okay, you're arguing for an *intelligent* evolution as opposed to an unintelligent and accidental one – but where's your evidence?"

'True. But those who argue that evolution is blind and *unintelligent* have no proof or evidence for their claims either. It's just a materialist article of faith. Take neo-Darwinism's hypothetical engines of evolution, allegedly random genetic mutations that supposedly accidentally arrived, survived, bred and thrived. Even if we agree that genetic mutations *are* the drivers of evolution it doesn't solve *where* the astonishingly clever *digitally encoded genetic instructions and information* that they contain came from, *in the first place* – the 'arrival of the fittest' is what we need to explain as much as the 'survival of the fittest' – as S. Meyer makes clear in his important book, *Signature in the Cell: DNA and the Evidence for Intelligent Design*.[108]

'Secondly, there is no way of proving that such mutations even if they have been the drivers of Life's evolution would have been random or accidental as opposed to *intelligent* and intended. It is just a *non-evidence-based* materialist belief or superstition that 'accidents-build-clever-cellular-machinery-and-codes'. The only rational way that genetic mutations can be seen as a driver of the unfoldment of the Living World, across deep time, is that if they were *intelligently* induced ones. For a deeper exploration of this you only need to read M. Behe's *The Edge of Evolution: The Search for the Limits of Darwinism*[109] where Behe explains that random genetic mutations are very limited in their 'creative' effects. He shows that they cannot explain the changes seen in the Life Forms over millions of years.

'In *Darwin's Black Box* Behe had already explained that many biological mechanisms, such as the blood clotting cascade and the bacterial flagellum are *irreducibly complex* which means that random, incremental Darwinian evolutionism cannot account for their 'all-or-nothing' nature. They either arise, just as many of the mechanisms that we make, all at once by some 'super' or higher Natural processes involving intelligence and directed purpose or not at all. In *The Edge of Evolution: The Search for the Limits of Darwinism* Behe adds that although Darwin's theory can explain minor evolutionary changes, random mutation and natural selection *explain very little* of the basic machinery of life. This is because most genetic mutations are harmful and constructive genetic mutations are so rare that they are vastly too infrequent to explain the Living World. This means that if genetic

---

[108] 2009. Harper One.
[109] 2008. The Free Press

mutations are one of the drivers of evolution they can only be *nonrandom* ones, that is ones that are intelligent in origin.

'In his book, *Darwinism: The Refutation of a Myth*, Soren Lovtrup explained that:

> '... the reasons for rejecting Darwin's proposal were many, but first of all that many *innovations* cannot possibly come into existence through the accumulation of many small steps, ... *natural selection cannot accomplish it,* because *incipient* and intermediate stages are *not advantageous* [so they would not be naturally selected].[110]

'One of Darwin's most formidable critics, making just this point was his contemporary, St. George Mivart. In *On the Genesis of Species,*[111] he explained that natural selection simply cannot account for the incipient or early stages of useful structures.'

'In other words, Ollie, they're saying that natural selection cannot *initiate* or create anything – it cannot account for the 'arrival' of the fittest but only for the survival of the fittest? It can only 'select' or preserve what already exists?

'Yes, exactly. Natural selection is not a 'creative' force. It simply describes allegedly random variations, of what already so cleverly exists. Further, the variations that do actually occur and which drive finch beak and peppered moth micro–evolution, are nothing more than reversible, backwards and forwards variations within existing gene pools, no different than the variations which intelligently designing human breeders have used, over time, to breed wolves into chihuahuas or Irish wolfhounds without ever giving rise to a genuinely new species. All dog breeds, tiny and giant, are still members of *Canis Lupus.* In other words they don't involve genetic mutations at all.

'All the evidence from artificial breeding, as Darwin knew, is that it is *never possible* to artificially breed a species beyond a certain limit without damaging or destroying the species in question. Extreme breeding of fruit flies, involving artificially induced genetic mutations, has never led to a truly new species ever arising but merely to existing ones being damaged. For example, as Jonathan Wells explains in *Icons of Evolution: Science or Myth? Why much of what we teach about evolution is wrong,* skillful geneticists have managed to artificially

---

[110] Lovtrup, S. 1987. *Darwinism: The Refutation of a Myth.* Springer.
[111] St. George Mivart. 1871. *On the Genesis of Species.*

breed fruit flies with an additional pair of wings but these fruit flies could never survive in the wild. They are seriously handicapped because the extra wings have no flight muscles. Wells points out that, contrary to Darwinian claims, this clever work provides:

> '... no evidence that DNA mutations supply the raw materials for morphological [that is macro, large-scale] evolution.[112]... As evidence for evolution, the four–winged fruit fly is no better than a two headed calf in a circus sideshow.'[113]

'Wells also says:

> 'Like fruit flies, human beings begin life as a single fertilized egg cell. As the egg divides, it bequeaths a full set of genes to each of its progeny. Eventually, the fertilized egg divides into several hundred types of cells: A skin cell is different from a muscle cell, which in turn is different from a nerve cell, and so on. Yet with a few exceptions, *all the cell types contain the same genes* as the fertilized egg.
>
> 'The presence of identical genes in cells that are radically different from each other is known as "genomic equivalence". For a neo-Darwinist, genomic equivalence is a paradox. If genes control development, and the genes in every cell are the same, why are the cells so different?
>
> 'According to the standard explanation, cells differ because the genes are differentially turned on or off. Cells in one part of the embryo turn on some genes, while cells in another part, others. This certainly happens ... But it doesn't resolve the paradox, because it means *the genes are being turned on or off by factors outside themselves*. In other words, control rests with something beyond the genes – something **epigenetic**. This [means] that genes are being regulated by cellular factors outside the DNA. ...
>
> '... neo-Darwinian genetics [has] never resolved the paradox of genomic equivalence. In fact, the paradox recently deepened with the discovery that developmental genes such as *Ultrabithorax* are similar in many different animals–including flies and humans. If our developmental genes are similar to those of other animals, why don't we give birth to fruit flies instead of human beings?'[114]

'The point of mentioning this, Ali, is that despite Darwinian assertions to the contrary *no one currently knows* what are the *true*

---

[112] Wells, J. 2002. *Icons of Evolution: Science or Myth?* Regnery Publishing, pp. 178.
[113] Ibid. p.188.
[114] Ibid. p.191–193.

mechanisms and drivers of Life's unfoldment or of its macro–evolution over millions of years. Once again, as Antony Latham points out:

> 'What is *not in any way explained*... are the sudden appearance of completely new forms and structures – [bacteria to Fish, Fish to Dinosaurs, Dinosaurs to Birds] indeed, all the appearances of the phyla [body plans] and classes (higher taxa). *These* **remain a total mystery and do not fit in with any known mechanism or theory**.[115]

'However, just because we don't have a precise mechanism or mechanisms for the Living World's roll out across vast time, does not mean that we don't have the intellectual tools to decide whether or not it is an *intelligent* as opposed to an accidental phenomenon.

'You mean scientists have the ability to decide whether it makes more sense to fill our current knowledge–gaps by reference to materialism's preferred causative agencies, its unreasonable gods of mindlessness, unintelligence and accidental chemistry or philosophical idealism's preferred causative agencies, ones involving extremely intelligent, higher–level or superordinate Natural laws, forces and processes regardless of their physical invisibility and their currently unknown nature?'

'Yes, exactly. After all, at one time we believed electricity and magnetism to be 'super' Natural forces and only decided they were 'ordinary' Natural ones when we understood their workings.

'So every time we understand some part of Nature it goes from being 'super' to 'ordinary'?'

'Yes, that has been the attitude. It's called 'naturalization'. It is based on the mistaken assumption, in materialist thinking, that to **describe** how something 'works' or what it's 'made of' (it's Aristotelian mechanical and material causes) is the same as fully **explaining** it. Materialist thought argues that gravity and electricity and magnetism are not 'super' Natural phenomena because of its belief that the existence of the entire universe is not a 'super' or 'higher' Natural phenomenon, it's merely an 'ordinary' Natural one – by which is meant that it is mindless, unintelligent and accidental.

'Ironically, materialism justifies this posture precisely as the scientific–Idealists and the religious and spiritual thinkers do, by saying:

---

[115] Latham, A. *The Naked Emperor: Darwinism Exposed*. London: Janus, p. 59. Emphasis and words in square brackets added.

"We don't know *why* the Big Bang or matter or the laws of light, gravity, electricity and the smart-materials-chemicals are as they are but *they just are* and it's all mindless, unintelligent and accidental or MUA!"

'The idealist and religious thinkers reply:

'We also don't know *why* the Big Bang or matter or the laws of light, gravity, electricity and the smart-materials-chemicals are as they are but, as you say, *they just are* and it's all an intelligent, purposeful, amazing, ultimate miracle, an AUM!

'We haven't a clue why there should be an intelligent and meaningful universe anymore than you know why, as you believe, there should be an unintelligent and accidental one. The only difference is that where you fill your knowledge–gaps with words like 'random', 'mindless' and 'accidental' we, taking our own immaterial minds and intelligences and the stunning, second by second, 24/7–cleverness of Nature's laws and Living Nature's mechanisms as clues, fill our knowledge–gaps with words like 'intelligence', 'purpose' and 'meaning'.

'Where you see a 'mindless' and 'accidental' universe, we see an intelligent and purposeful one. You seem to think that by explaining the clever workings of electricity, gravity, DNA or fruit flies that you have explained *why* these are whereas you have simply described their workings, their material and mechanical causes, and whatever you currently do not understand about them, such as their ultimate origins, you put it down to unintelligence and *'accident'* where we opt for purpose and *'intelligence'*. We see total Holistic Reality as intelligent and purposeful, you see it as mindless and meaningless. You see it as the job of science to intelligently explain how *unintelligent, purposeless and meaningless* it all is. We see it as the job of science to intelligently discriminate between those features of Nature that are unintelligent and random and those that are intelligent and purposeful.'

'It is because the materialist world–view which predominates in mainstream science today hypothesizes that existence as a whole is utterly mindless, unintelligent and 'accidental' or MUA that it *has to insist,* to stay within its logic, that the Living World 'must be' an accident too. It cannot stay within the logic, however, because if the Cosmos really was 'mindless and unintelligent' it could *not* give birth to *minds* and to *intelligence* – including the immaterial minds and intelligences of those who see it as the intelligent task of science to intelligently explain how *mindless and unintelligent* the Universe, including Life on Earth is.

'Idealism, by contrast, suffers under no such self-refuting self-constraints. It keeps an open mind and explains *unintelligent* and accidental effects, like randomly evolved river valleys as having (relatively speaking) unintelligent and accidental causes and *intelligent* effects (like Rolls-Royces, Jumbo Jets, DNA CODE, "more advanced than any software", tiny Living Cells, "even more complex than Jumbo Jets", fruit flies, butterflies, blue whales and people) as having *intelligent* causes.

'Look, desert islanders faced with an unknown Rolls-Royce that appeared out of the blue on their island one day, could perfectly rationally conclude that the magnificent machine is *intelligent* in origin, based on all kinds of criteria (irreducible complexity, specific complexity, logic, commonsense, etc, etc). The rationality of their conclusion is *not* dependent on whether they know precisely *how* the magnificent car was, through time, intelligently designed or *intelligently* evolved.

'So, equally, we may not know the precise mechanisms of the *intelligent* unfoldment of the Living World, across vast time, but that is not evidence for, let alone confirmation of the materialist scientific hypothesis that they 'must be' unintelligent and 'accidental' then ... especially as *all* our real-world knowledge of the differences between *intelligent* and *accidental* causation and between guided chemistry and accidental chemistry point implacably towards *intelligence,* whatever the precise nature of its constituent higher–order laws, forces and processes, as not only the better but the *only* reasonable explanation of the Living World's ultimate sources and causes, quite regardless of their physical invisibility, moral merits or mysterious purposes, than towards mindlessness and accident.'

'So we live in a clever, weird and wonderful universe then, Ollie, not a mindless and accidental one.'

'Yes, of course we do. And, while it's true that the knowledge that Life and the Living World are incredibly intelligent phenomena may make the universe seem 'spooky' or 'weird' to some people, just as the modern and steadily accumulating evidence for Life after death may seem 'spooky' or 'weird' to some, at least such conclusions match the data of reality. After all, isn't *Life!* weird and wonderful and thoroughly *mysterious!* Those who argue otherwise are not paying attention.'

We walked on.

8 People: Accidental 'Bio-Robot Computers' or Living Souls?

The Darwinian and other materialist theories of 'random' evolution maintain that humans are 'accidental' bio–robot computers run by 'accidental' digital DNA machine CODING, 'more advanced than any software ever created'. Is this view scientifically justified? In the real world, robots and their guiding software codes never arise 'by accident'.

'After Darwin published *On the Origin of Species*, in 1859, and the dramatic implications of his ideas began to sink in, many people came to believe, Ali, that the miracle of very Life itself was an 'accident' – and the entire Universe, from atom to galaxy, fairly rapidly went from being understood as a magical, multi-dimensional and *intelligent* Mystery all the way up and all the way down, to a meaningless and *'accidental'* Mystery, all the way up and all the way down. Some people even started to think of humans as Soul-less, 'accidental', Darwinian bio-robots.'

'People are not Soulless 'bio-robots', for heaven's sake!'

'You say that, but that is the logic of the materialist scheme because if *only* random chemicals + random events exist, then you and I and everyone else and all the beautiful plants and animals are, by definition, 'random events' too.'

'But it's ridiculous.'

'I agree. It is, however, the inevitable conclusion of Darwinian materialism. And, although we are not 'robots' in that, unlike robots, we have independent minds, Souls and free wills, we are, in a sense, *Soulful* bio-robots in that each human 'bio-robot' body is, more or less, like all the others. It is only the inhabiting minds and Souls that display great differences in intelligence, talent, character, virtue and vice. The body of an Aristotle, Cleopatra, Caesar, Mary, Jesus, Joan of Arc, Marie de Medici, Bach, Mozart, Napoleon, Marie Curie or Einstein does not look that different from anyone else's; it is their hearts, minds and Souls that express vastly different levels of evolutionary Soul development and creative potential.

'I mentioned the 'bio-robot' analogy because, in materialist and Darwinian unintelligent design thinking humans are often likened to fabulously complex bio-robots or super-computers with heads, faces, arms and legs.'

'But robots and super–computers are never 'unintelligently designed', they *never* arise by 'accidental' evolutionary processes.'

'That's true. Our universal, empirical experience is that they are always *intelligently* evolved and intelligently designed and the digital computer codes, machine languages and software that run them (far simpler than the digital DNA CODING helping to run our bodies), are also always intelligently designed and evolved. In fact the DNA CODING running our bodies is far more advanced than any software we use to drive any robots we make.'

'One of the problems we have in making the idealist or non-materialist case for intelligent design, intelligent evolution and a rational, multi-dimensional Cosmos of intelligence and meaning is that modern materialistic science has taken far too much notice of the philosopher Hume's argument against analogy. "Yes," the Humean argument goes, "watches, cars and computers don't arise by us randomly throwing the parts around, or tumbling them in barrels labeled 'Watches', 'Cars', 'TVs' and 'Computers', but the Life Forms are nothing like watches, cars, TVs and computers, so the analogy argument fails. This means that accidents 'must be' creative then, mustn't they? What other choice is there in a meaningless, unintelligent and accidental Cosmic Mystery of Existence?"

'This, however, is philosophically and scientifically unjustified. Firstly, it assumes the very thing that is in question and which it has not proved – that the Cosmos is meaningless, unintelligent and accidental. Secondly, it's blazingly obvious that 'Life is more than chemistry' – basic, I know, but this obvious point seems to escape mainstream scientific culture due, perhaps, to the artificial synthesis of urea in 1828 and the strange refusal to acknowledge the existence of the Life Principle and the Life or Vital Forces as they used to be called. Thirdly, the same ideologically materialist thinking that uses Hume in this way this doesn't hesitate, when it suits it to do so, to think analogically and to conceive of humans as 'accidental bio-robots' with planet-sized supercomputer brains that supposedly arose and arise right now, clever second by clever second, 24/7, by allegedly entirely unintelligent, non-teleological and 'accidental' processes, even though, in the real world, computers and their driving codes never arise 'by accident', let alone mechanisms like Living Cells, more complex than supercomputers, that can reproduce themselves in a matter of hours! Fourthly, it's true that the Life forms are not like any machinery we make. This is because they are orders of magnitude more complex. That strengthens the machine analogy argument it hardly weakens it.

'Finally, some of the key modern arguments for intelligent design in

Living Nature are *not*, in any case, analogical arguments. It is **not an analogy** to say that the Life Forms are to a vital extent run by the digital DNA CODING that, like all software and codes, for mathematical and statistical reasons could not possibly arise 'by accident'. This is because the digital DNA Coding is **not *'like a code'*,** which would be an argument from analogy, ***it is a code!***[116] And, as information systems theorist Werner Gitt says:

> 'The basic flaw of all [materialistic] evolutionary views is the origin of the information in living beings. *It has never been shown that a coding system and semantic [meaningful] information could originate by itself… The information theorems predict that this will never be possible.*'[117]

'We work very hard and *intelligently* to build robots and computers and to *intelligently* program them with artificial *intelligence* not with 'unintelligence' – we don't tumble their parts in barrels labeled (1) 'Robot', (2) 'On-Board Computer' and (3) 'digital CODING Software' and hope that after a sufficiently long period of tumbling, millions of years in the case of the Life Forms, that the completed products will emerge, 'by accident'. Yet, mostly, it seems, because we cannot *physically* see Living Nature's stunningly intelligent, super or higher Natural laws, forces and causes, because we are reluctant to *rationally infer* them, because we are unwilling to take any notice of collective humanity's intuitive witness and testimony to their existence and in total conflict with all our real–world experience of the 'creative' and 'evolutionary' causal capacities of 'accidental', Life Forceless, mineral chemistry, we consider it to be 'rational' to imagine ourselves to be Living!, breathing!, breeding!, 'accidental–robots' or 'accidental–super–computers-on-legs' with Universal Witnessing Awareness, Sentiency, Subjectivity, Feelings and Emotions, Hearts and Minds, Ideas, Thoughts, Intuitions, Visions, Inspirations, Imaginations, Dreams and Plans… These are all complete and utter self-refuting contradictions.

'The only reason to imagine that Living Cells more complex than Space Shuttles and Human Beings with vast 'super–computer' sized brains each, every single one, with more molecular-scale switches than *all* the computers, routers and internet connections on the entire

---

[116] Meyer, S. 2009. *Signature in the Cell.* Harper One. Gitt, W. 2006. *In the Beginning Was Information.* Master Books, p. 124.

[117] Gitt, W. 2006. *In the Beginning Was Information.* Master Books, p. 124. Emphasis and text in square brackets added.

planet,[118] are the unintended and 'accidental' results of *Non-Living* Amino acids bumping into each other is if we have an unquestioning (and unreasonable, because it is not evidence-based) faith and belief in the creative powers of accidents of chemistry. As Michael Ruse says:

> "evolution has functioned as something with elements which are, let us say, akin to being *a secular religion*" (Ruse 1993)[119]

'Yet, where is the extraordinary evidence for such extraordinary faith claims? Where is the evidence that 'accidental chemistry' is 'creative' or 'evolutionary'? There is *no empirical evidence* for this unreasonable hallucination that has afflicted western thinking for the last one hundred and fifty Darwinian years.

'But the materialist argument, Ollie, is *why* should the Cosmos or All that Is be intelligent? Why shouldn't it be numb, 'dumb', mindless, meaningless and pointless as the materialists believe? Why shouldn't it just be one big 'accident'? After all there sometimes seems to be little rhyme or reason to our brief lives and Nature can be incredibly harsh?'

'Ali, I know that. As I said before, I'm not trying to make an argument for the perfection of existence, but for its *intelligence*. Stealth Bombers are neither technically perfect nor morally perfect devices but it takes *huge intelligence* to design and evolve them through time and they function *intelligently* – they don't arise by means of the great but impossible materialist engines of pseudo 'creativity': Mindlessness, Unintelligence and Accident or MUA.

'So the universe contains some pretty peculiar intelligences then?

'Look, it may do. That, however, is a vastly more rational argument than claiming that it contains no intelligence at all. The point is, Ali, no one knows *why* anything exists – *why* there should be a universe rather than not, *why* there should be even one photon rather than not.'

---

[118] Moore, E. A. 11.17.2010. cnet. "In the cerebral cortex alone, there are roughly 125 trillion synapses, which is about how many stars fill 1,500 Milky Way galaxies. ... Researchers at the Stanford University School of Medicine ... [have] found that the brain's complexity is beyond anything they'd imagined, almost to the point of being beyond belief, says Stephen Smith, a professor of molecular and cellular physiology ... One synapse, by itself, is more like a microprocessor ... In fact, one synapse may contain on the order of 1,000 molecular-scale switches. A single human brain has more switches than all the computers and routers and Internet connections on Earth."

[119] Battson, A. 1997. *Facts, Fossils, and Philosophy*. Author's website. Emphasis added.

'But don't they know that there was a Big Bang etc?'

'Yes, they do but that's no different to desert islanders saying: "Yesterday there was no Rolls-Royce on our island. Today, who knows why, there is. So let's try and work out how it works and how to use it."

'I don't quite see the parallel?'

'Look, the desert islanders don't know *why* a Rolls-Royce suddenly appeared, while they slept, in the dead of Night on their island (any more than we know *why* the Big Bang happened) or what it's for. All the islanders can ask is whether the shiny Car (or the Big Bang or the Living World) has a meaning or purpose or is it meaningless? All they can work out is (a) how to use it (as our sciences and technologies do in respect of the Living World and Nature in general) and (b) whether it makes *more sense, based on all that they know* of different kinds of causation – *intelligent,* anti–entropic causation (mechanisms and devices that the islanders themselves make) and *accidental,* entropic causation (randomly evolved river valleys, rubbly screed, wave marked patterns on sandy beaches) – to conclude that it's a device created and evolved by (a) *intelligent, meaningful and purposeful* sources and causes or (b) one created by materialism's impotent, irrational, pseudo-uber-gods of Mindlessness, Unintelligence and Accident or MUA.

'The trouble is, Alisha, that, after Darwin, science adopted the stance of methodological–materialism and really began to think and insist that it would be possible to ignore the last two of Aristotle's famous four causes – material ('made of'), mechanical ('how it works'), formal (shaping ideas), and final (purposes) – and to be able to explain the origins and evolution of the TVs (the Life Forms in the metaphor) solely in terms of (1) mindless matter – just metallic ores and oils that 'accidentally' mined and refined themselves out of the ground, 'accidentally' became wires and circuit boards in the TVs and (2) purposeless, unintelligent mechanism – just wires, plastics and circuit boards that 'accidentally' self-assembled and became TVs), on the unrealistic hypothesis and unjustified assumption that they *had* and *have* no (3) intelligent, Aristotelian, shaping (formal) causes (engineers and technologists) or (4) intelligent purposes (final causes – the transmission of TV Shows, (Souls)) behind them. This, however, doesn't work. It simply leads to intellectual absurdities.

We both fell quiet. After continuing on for a while in the silence broken only by the rustic sounds of the animals in the fields and the murmuring of the river Ali looked at me thoughtfully and said:

'One thing, Ollie, until you told me a little bit about this today, I had no idea that Living Cells were such incredibly complex mechanisms in their own right and that there is no realistic way that they could ever arise by random means.

'Yes, it's true:

> 'the probability of life originating at random is so utterly miniscule as to make it absurd ... It ... must reflect ... *higher intelligences* ... such a theory is so obvious that one wonders why it is not widely accepted as being *self-evident*. The reasons are *psychological* rather than scientific.'[120]

'This is why I care about this Ali. I think science is amazing but exclusively materialist thinking has turned it into something unreasonable, like a 'church' that leads to darkness not enlightenment. A de-intelligencing and de-Souling 'church of science' that does not seek to confirm the true Life, Soul and *intelligence* in all things but puts a lot of effort into denying them. A faustian science some of whose advocates proclaim that God, 'super' Nature or supreme intelligent Beingness are illusions; which maintains, with no evidence but its own beliefs, that we live within an unintelligent and mono-dimensional Cosmic Mystery rather than a blazingly intelligent, Alive and Soulful Cosmic Mystery, a science which holds that humanity's greatest thinkers and teachers were all wrong, that humanity as a whole is wrong, that the Laws of Entropy that prevent the 'accidental' creation of intelligent 'machines' of any kind, let lone the clever codes to run them, are wrong and that the Living World is a second by second manifestation of mindlessness, unintelligence and accident – that it is a billions year long running *'accidental chemical reaction'*.

'It's not, Ali, that I think that existence is all perfect and wonderful or that the world's religions and spiritual traditions understand and explain reality perfectly, far from it. But to persist with the materialist hypothesis, as the supposedly 'majority' view in science and biology, that the universe as a whole and the Living World in particular are meaningless 'coincidences' just because their ultimate or final intelligent sources and causes cannot be *physically* seen, and to continue to do so in conflict with the ocean of human testimony, all the mystical, psychical and spiritual experiences confirming the *rational scientific inferences,* logical philosophical deductions and the spiritual insights of almost all the greatest minds and thinkers prior to Darwin and many since – Buddha, Lao Tzu, Socrates, Plato, Aristotle, Jesus, Plotinus, Muhammed, Avicenna, Aquinas, Bacon, Galileo, Newton, Boyle,

---

[120] Hoyle, F. and N.C. 1981. Wickramasinghe, *Evolution from Space*. J. M. Dent & Sons.

Faraday, Maxwell, Planck, Heisenberg and countless others – is unreasonable and a view well past its sell by date.

'A view which, at its extremes, insists that we do not have any true minds at all, we are 'unconscious zombies', as Dennett put it, 'accidental robots' (a self-contradiction) or 'accidental computers' made of 'meat' as someone called Minski once crudely put it,[121] never mind that no computer of any kind has ever arisen 'by chance'. How can our civilization have become so irrational? And, tiny as they are, even the humble Living Cells at the base and cause of all the larger Life Forms, and here from the very beginning, billions of years ago, with no evidence anywhere for having arisen by random Darwinian style processes,[122] are the *most* dynamically complex *single* mechanisms *within* the larger Life Forms, far more complex than the larger organs like stomachs, lungs, intestines, eyes, noses, etc, that they help to build.

'This is what a leading cell biologist says:

'We have always underestimated cells. ... The entire cell can be viewed as a factory that contains an elaborate network of interlocking assembly lines, each of which is composed of a set of large protein machines. ... Why do we call the large protein assemblies that underlie cell function protein *machines*? Precisely because, like machines invented by humans to deal efficiently with the macroscopic world, these protein assemblies contain highly coordinated moving parts.'[123]

'The crucial point is, Ali, that nothing in humanity's experience – the physical invisibility of the intelligently causal dimensions of total Holistic Natural Reality included – gives us reasonable scientific grounds for believing that such intracellular machines and machinery could ever arise *'by accident'* or be arising and functioning, right now, 24/7, 7/52, clever microsecond by clever microsecond by Life–Forceless and unintelligent processes.

---

[121] Minsky M. Quoted in: Michalowski S. "Science, Man and the International Year of Physics." OECD Global Science Forum. Quoted by L. Dossey: 2010. *Is the Universe Merely A Statistical Accident? The Blog. Huffington Post.*

[122] Denton, M. 1985. *Evolution: A Theory in Crisis.* Burnett Books. Behe, M. 1996. *Darwin's Black Box.* Free Press.

[123] Bruce Alberts (former president of the National Academy of Sciences), 1998. 'The cell as a collection of protein machines: Preparing the next generation of molecular biologists'. *Cell, 92* (February 8, 1998): 291.

'It is only materialist thinkers in some kind of war with the 'super' Natural or with the idea that total Nature is *intelligent* and *meaningful* as opposed to 'mindless' and 'accidental' who have to so surreally insist, in a kind of perverse, Naked Emperor in reverse way, that the Living World may *'look' very finely dressed in intelligence, beauty, meaning and purpose but: "It isn't really!"* This is like saying that the TV or super-computer may 'look' like a very 'intelligent' device but it isn't really, no intelligence went into its arising, it must be an 'accident' of some kind.

'Materialist dominated science and biology have taken our culture into these realms of unreasonableness simply to avoid the classical rational inference that our *universal witnessing consciousness, our sentiency, intelligence and purposefulness,* our capacities for love, sensibility, art, abstract thought, mathematics and philosophy and the incredibly *smart,* 24/7–dynamic mechanisms of the Living World all arise from a blazingly *intelligent and purposeful* Cosmos, no matter how weird and mysterious, than from a 'mindless' and 'accidental' one which is just as weird and mysterious but is supposedly 'utterly unintelligent' and utterly 'meaningless' into the bargain!

'After all, Ali, why is it supposedly (a) *more* reasonable, as materialist thought holds, that there should be an utterly mindless, unintelligent and accidental Cosmic Mystery that's ***just meaninglessly there*** or which suddenly appeared, from no–where to now–here, in a stupendous Big Bang, for no reason that anyone can see, than that (b), as classical Idealist and traditional religious and spiritual thought holds, there is a blazingly *intelligent* and *purposeful* Cosmic Mystery that's ***just brilliantly and meaningfully there***.

'Which is the better way of explaining the data of reality, Ollie? A mindless and meaningless 'weird' Cosmic Mystery or a brilliant and intelligent 'weird' Cosmic Mystery!'

'That's it. Those, in essence, are our only two choices. We cannot escape the ultimate mystery of it all but we can decide which kind of amazing Cosmic Mystery we think we're in – based on the data of human experience consisting in: (a) so-called 'ordinary' sensory data and (b) higher–sensory, 'higher-frequency' data subjected, in both cases, to the light of reason, intuition, rational inferences and logical deductions. I know which of the two views makes more sense to me.

'Just because we haven't a clue, right now, from the perspective of our conventional physical–sciences how or why intelligent laws, forces and processes in a multi-dimensional Cosmos of Intelligence and meaning, can or would (a) give rise to incredibly smart–materials–

chemicals in the first place (b) give rise to super-ordinate Life and Soul Forces (c) express and evolve amazing Soul Life Form Beings whose ultimate meanings and purposes we don't yet even begin to understand – the fabulously magical Plants, Animals and People – doesn't mean that a multi-dimensional, multi-vibrational Cosmos of *Life and Soul and purpose and intelligence,* (rather than one of total mindlessness and accident), is not the more *reasonable* description of Reality, than to continue using our very own mysterious consciousnesses, *minds and intelligences, all immaterial* qualities, all non–physical *essences* of Soul and Spirit, to deny that we or the astonishingly clever mechanisms of Living Nature or the Universe at large have or embody any of these at all and that, at the extremes of materialist thought we don't even really have minds, that:

> "We're all zombies. Nobody is conscious."[124]

This way of thinking, it seems to me, Alisha, is intellectually unreasonable, a delusion or collusion that children and young people are so Soul destroyingly and relentlessly exposed to, decade after decade, and we wonder why they are often so disillusioned with Life and existence. Dr. Larry Dossey puts it very well.[125] He says:

> 'There's an even drearier little secret that veteran scientists never let kids in on – that if they enter science, they have to check their minds at the door. The reason is that ***mind,*** as most people think about it, ***does not exist*** in conventional science, because the expressions of consciousness, such as choice, will, emotions, and even logic are said to be *brain in disguise.* ... Nobelist Francis Crick in his 1995 book *The Astonishing Hypothesis* [said] 'You, your joys and your sorrows, your memories and free will, are in fact no more than the behavior of a vast assembly of nerve cells and their associated molecules. As Lewis Carroll's Alice might have phrased it: "You're nothing but a pack of neurons."'[126] Or, as Marvin Minsky, the Massachusetts Institute of Technology cognitive scientist and artificial intelligence expert, put it more crudely, "The brain is just a computer made of meat."[127]

---

[124] Dennett D. 1992. *Consciousness Explained.* Back Bay Books, p. 406.

[125] Dossey, L. 2010. Is the Universe Merely A Statistical Accident? *The Blog. Huffington Post.*

[126] Crick F. 1995. *The Astonishing Hypothesis.* Scribner; p. 3.

[127] Minsky M. Quoted in: Michalowski S. "Science, Man and the International Year of Physics." OECD Global Science Forum.

'Do you see how unreasonable and self-refuting these materialist statements are? Ones made by clever, even brilliant minds. Do computers of any kind – let alone ones 'made of meat' – ever arise by accident? Could *any* kind of computer, anywhere, at any time, arise 'by accident'? The materialist answer to that question is, in my view, simple, simplistic and absurd. It goes like this:

> 'Because we can't physically see Life's, subtle, intelligent 'higher' Natural Sources and Causes (and are unwilling to *rationally infer* them or to take any notice at all of collective humanity's intuitions and spiritual and mystical experiences that tell us they are real), we believe the universe must be an 'accident'. Therefore, it logically follows that Life on Earth and our own *minds and intelligences* must be 'accidents' too! So computers 'made of meat' *must* arise by 'accident. That's just logic!'

'The argument may be logical but the premise is false and a false premise generally leads to a false conclusion. The false premise is that the physical invisibility of Life's intelligently causal or 'super' or Higher Natural sources is 'proof' or evidence that the Cosmos and we who are its children must all be 'accidents'. Yet, intelligent causation can be rationally inferred or intuited even if it cannot be physically seen.

'Larry Dossey continues:

> 'Crick went further. In his subsequent book *Of Molecules and Men*, he wrote, "The ultimate aim of the modern movement in biology is to explain all biology in terms of physics and chemistry"[128] – to analyze, in other words, the meat. And lest there be no doubt about where he stands, philosopher Dennett says, "We're all zombies. Nobody is conscious."[129]

'Wow, Ollie, it is ridiculous . . .'
'This is what I'm saying. The philosophy of materialism, of which Darwinism is the keystone in the whole de-intelligencing, de-Souling materialist arch, has made both science and philosophy illogical and unreasonable. It is as though Mephistopheles himself is standing in the wings of history laughing at us as we, Faust's offspring, use our very amazing and utterly immaterial *minds, Souls and intelligences* to deny that we or the Living World and the Cosmos as a whole have any of

---

[128] Crick F. 2004. *Of Molecules and Men*. Prometheus, p. 10.
[129] Dennett D. 1992. *Consciousness Explained*. Back Bay Books, p. 406.

these – to insist, instead, that we are 'accidental' computers 'made of meat'.

'Larry Dossey says:

> 'Try selling that to a teenager contemplating a career in science and see what happens.
>
> 'Novelist Arthur Koestler poked fun at these positions by taking aim at Rene Descartes, the seventeenth-century philosopher who was extraordinarily influential in establishing the notion of a mindless body. "If ... Descartes ... had kept a poodle, the history of philosophy would have been different," Koestler wrote. "The poodle would have taught Descartes that contrary to his doctrine, animals are not machines, and hence the human body is not a machine, forever separated from the mind ..."[130]
>
> 'This morose, meaningless side of science is never openly presented to young students contemplating a lifetime in science. They usually sniff it out later on, after a career choice has been made. I know of no studies that assess the impact of these dark views on young scientists when they encounter them. Are they negatively affected? Do they adopt a chin-up attitude and soldier on, having traveled too far on the science path to turn around? Or – most commonly, I believe – do they schizophrenically partition their psychological, spiritual and scientific lives into separate domains in a desperate attempt to find balance, silently suffering the jagged contradictions the rest of their life?
>
> 'Purists insist that science is neutral on matters of meaning; the world is what it is. Whatever meaning we find in the world comes from us, not the world itself. We read meaning into the world, not from it. This sword cuts two ways; *if meaning should not be imputed to the universe, neither should meaninglessness.* It is a plain fact that scientists in general, peering into the *same* universe and aware of the *same* set of facts, see meaning in different ways, ways that are not part of science itself. No [physical–]scientist has ever possessed a meaning meter. Therefore the proper approach, it would seem, would be to declare *questions of meaning beyond the purview of [physical–]science* and to cease imposing one's personal view as the official way the universe should be interpreted. This would give students and young scientists a fighting chance to find their own path where meaning and purpose are concerned, and not be bullied by senior scientists who ought to know better.' [131]

---

[130] Koestler, A. 1978. *Janus: A Summing Up.* Random House, p. 229.
[131] Dossey, L. 2010. Is the Universe Merely A Statistical Accident? *The Blog. Huffington Post.* Emphasis added.

'Raymond Tallis, a neuroscientist and ontological agnostic, a non-religious thinker, has written a devastating critique, *Aping Mankind,* of the modern attempt to reduce people to 'brains', that is to 'computers made of meat' and nothing more:

> '[He fearlessly criticizes] the exaggerated claims made for the ability of neuroscience and evolutionary theory to explain human consciousness, behavior, culture and society. ... Tallis directs his guns at neuroscience's dark companion – Neuromania, as he describes it ... Tallis dismantles the idea that "we are our brains", which has given rise to a plethora of neuro-prefixed pseudo-disciplines ... shows it to be confused and fallacious, and an abuse of the prestige of science, one that sidesteps a whole range of mind-body problems. ... offering a grotesquely simplified and degrading account of humanity.'[132]

'In *Irreducible Mind: Toward a Psychology for the 21st Century* the authors challenge the:

> 'Current mainstream opinion in psychology, neuroscience, and philosophy of mind [that] holds that all aspects of human mind and consciousness are generated by physical processes occurring in brains. ... this **reductive materialism is** not only incomplete but **false**. The authors marshal evidence for a variety of psychological phenomena that are extremely difficult, and in some cases clearly **impossible, to account for in conventional physicalist terms.** Topics addressed include... memory, psychological automatisms and secondary personality, near-death experiences and allied phenomena, genius-level creativity, and 'mystical' states of consciousness both spontaneous and drug-induced.
>
> 'The authors show that these phenomena are more readily accommodated by an alternative 'transmission' or 'filter' theory of mind/brain relations advanced over a century ago by a largely forgotten genius, F. W. H. Myers, and developed further by his friend and colleague William James. This theory ratifies the commonsense conception of human beings as causally effective conscious agents, [versus 'accidental' automata] and is fully compatible with leading-edge physics and neuroscience.'[133]

---

[132] Tallis, R. 2011. *Aping Mankind: Neuromania, Darwinitis and the Misrepresentation of Humanity.* Acumen Publishing; product description, Amazon.

[133] Kelly, E. F. & E. W. Kelly, A. Crabtree, A. Gauld 2009 *Irreducible Mind: Toward a Psychology for the 21st Century.* Rowman and Littlefield; product description, text in square brackets added.

'This ties in with what Larry Dossey is saying, Ollie, it makes sense.'

'Yes. I think though, that it is good to make a distinction between what physical–science can and cannot tell us and what spiritual–science can and cannot tell us.'

'How do you mean?'

'I mean that right now our knowledge of the world is split into two domains or 'magisteria' as the late S.J. Gould referred to them. Those of physical–science and those of spiritual–science. The first deals with the matter and mechanism or the 'tails' dimensions of existence (covering Aristotle's *material* and *mechanical causes*) and the second with the Life, mind, intelligence, purpose, Soul and subjective or 'heads' dimensions of existence (covering Aristotle's *formal* and *final causes,* the shaping ideas and purposes of things).

'What methodologically–materialist thinkers like Crick and Dennett are saying is that the second realm, (the 'heads' dimensions of Aristotle's formal and final causes), the *Conscious, Subjective, Life, Mind, Purpose, Love and Soul realms* of our multi–dimensional, multi–frequency, Holistic Natural Reality do not truly exist. They are merely the 'chance effects' of the kind of numb, dumb, solid, stolid, tiny atomic billiard ball 'matter' that materialism believes in – although this itself is a misguided view of 'm—a—t—t—e—r' which, as the Yogis of India always said, using highly trained, higher–sensory perception, and as modern quantum physics has confirmed, is not truly 'solid' at all but is made of intelligently in-form-ed quantum–energy s—p—a—c—e, mind-stuff or 'chitta' as the Yogis call it. A Cosmic Illusion, a Yogic–Maya, the whole a vast 'trick' or miracle of Consciousness and, ultimately, not 'material' at all. As the eminent physicist, Sir James Jeans said:

> '... the Universe begins to look more like a great thought than like a great machine. *Mind* no longer appears to be an accidental intruder into the realm of matter ... we ought rather hail it as the creator and governor of the realm of matter.[134]

'And Anthony Peake:

> 'Self-awareness within *a seemingly unaware universe simply does not make sense,* and psychic phenomena similarly present anomalies that may be the pointers to a hitherto unknown model

---

[134] J. Jeans, *The Mysterious Universe.* Cambridge University Press, 1931, p. 137.

of science; *a model in which mind creates matter rather than matter creates mind.*'[135]

'OK, Ollie, so if we really are 'bio–robots', 'computers made of meat' as modern Darwinism and methodologically materialist biology both maintain, then they cannot have it both ways. They should concede, if logic counts for anything, that we, like all robots and computers, always and everywhere, are *intelligently* designed or *intelligently* evolved bio–robots or DNA CODED 'super–computers–on–legs.''

'Ali, they can't admit to that though.'

'Why not?'

'Because admitting it *refutes materialism*. This is why this obvious contradiction in materialist thinking is quietly overlooked. Obviously, in the real world, outside topsy-turvy Darwinia, 'robots' (never mind bio-robots of the most staggeringly complex kind) *never,* ever arise or evolve 'by accident' – they are always *intelligently* designed and *intelligently* evolved.

'But how, Ollie? That's the problem for materialist thinking isn't it?'

'Yes, it's true. But that's the whole point.'

'How do you mean?'

'I mean the fact that we Live within a Mystery. Quite clearly it's irrational to say that some kind of bearded-god 'did it' and, more to the point, 'maintains and sustains it all, right now, dynamic second by dynamic second but it's every bit as irrational to claim, as materialist thought does, that its unreasonable and anti-empirical gods of Mindlessness, Unintelligence and Accidental Chemistry are responsible for Living Nature's 24/7–clever functioning.'

'So we are no further on then, are we?'

'No, that's not actually true. The task ahead is to develop and evolve our understanding of *intelligent* causality within and of Living Nature – that is of 'super' or Higher Natural causation – not waste more academic time claiming that *Life-Forceless, Soulless, 'accidental chemical reactions'* can get us to amazing human minds and the incredible Living World. Once we accept that we live within a mysterious multi–dimensional Reality which is, if you like, *very Life! itself,* which is itself *Alive!*, which is itself a vast, *Intelligent, Conscious, Sentient, Living, Cosmos of extending Beingness* we can begin to intellectually and spiritually move forward once more.

---

[135] Peake, A. *The Out of Body Experience: The History and Science of Astral Travel.* Watkins Publishing.

9 Court Says: Intelligent Design 'May Be True'

When 2000 years of western thinking and rationalism is overturned and to argue that Life and the Living World are 'accidents' is considered to be 'scientific' but to argue that that is not merely unlikely, it's a mathematical and statistical *impossibility,* is judged to be 'unscientific'.

'Once, Ali, we accept that we live within a mysterious multi–dimensional Reality which *is very Life* itself, which is Itself a vast, *Intelligent, Living, Cosmos of extending Beingness* we can begin, intellectually, emotionally and spiritually to move forward again. Then, all the palpable *intelligence and purposefulness* of our beautiful, baffling, challenging, magical and Soulful Living World all begin to make so much more sense.

'No one can say *why* this is the way things are, *why* we live within an *intelligent* Cosmic Mystery rather than a mindless and accidental one, let alone *why* there should be a Cosmos *at all, (mindless or intelligent)* but, as the Yogis pointed out centuries ago and some of the modern quantum physicists, like Max Planck, Sir James Jeans and Amit Goswami[136] have confirmed, that is just the way it is. We live and move and breathe in a Living, multi-dimensional, multi-level Cosmos of Soulful Meaning and Intelligence, not Soulless unconsciousness and accident.'

'Yet, those non–materialist thinkers, today, who argue that the most complex biological mechanisms known to humanity – we ourselves – could *never* have arisen, nor continue to arise right now, 24/7, 7/52, blazingly clever second by blazingly clever second, by unintelligent and random processes, and that it is possible to demonstrate this scientifically, philosophically, mathematically and statistically *are*, in some cases, even legally prevented from doing so, as, for example, in the controversial Dover court case in 2005. This is because to claim that *science,* not religion, can show that the Living World is not, after all, a billions year long running *'accidental chemical reaction'*, that it is scientifically and certifiably *impossible* for it to have arisen in such a

---

[136] Goswami, A. 1993. *The Self-Aware Universe: How Consciousness Creates the Material World.* Jeremy P. Tarcher.

way and, as important, for it to be continuing to arise, *right now, 24/7, 7/52* is judged, by some, to be not science but 'religion' and to be intellectually disreputable:

> 'No, Mr Galileo, your claim that the Earth orbits the sun and not the other way round is not *science* but *religion.*"

'Very funny, Ollie.'

'Actually, that's not being as absurd as you think. Galileo's discovery did have religious implications at the time which many thinkers did not like. Contrary to myth, it was *not* that the Earth ceased to be 'important' in his scheme – quite the reverse – it was because the medievals regarded the Earth as the *lowest* part of the creation, as the 'sump' of the universe, as its very 'drain', that they felt it should be at the centre and not elevated to the celestial levels, the heavenly spheres as they saw them, as it would be if Galileo's ideas were adopted.

'In one of the ironies of the Dover case, in 2005, as to whether modern intelligent design or ID theory could even be mentioned in class – not taught, just mentioned as an alternative view to Darwin's materialist theory and upon which there were books in the school library that interested students could look at, in their own time – the judge ruled that while intelligent design, or *ID,* arguments ***"may be true,*** a proposition on which the court takes no position, ID is not science".'

'What was his reason?'

'It was partly because some of those favoring, methodological–materialism ('accidents-build-clever-machinery-and-codes'), and unintelligent design thinking in science argued that classical, idealist ('intelligence-before-clever-mechanisms-and-codes'), intelligent design thinking was not science but religion.'

'Interesting, Ollie. So it is quite similar to what happened to poor Galileo. Given Michael Ruse's assessment of Darwinian evolutionism as "akin to being *a secular religion"* I suppose it could be argued that the Darwinian theory of unintelligent design shouldn't be taught in science class either!'

'I agree, and, contrary to myth, it was *not initially* the church but Galileo's own academic peers, the conservative consensus–science opponents of his day, equivalent to today's Darwinian and materialist theorists of unintelligent design, who went to the church authorities and said, "Look, heliocentrism is just *not science*, so Galileo must be banned from teaching his theories." In fact, Galileo avoided publicly stating his support for the heliocentric system of Copernicus for a very long time

Court Says: Intelligent Design 'May Be True'

because he was afraid, contrary to myth, not, initially, of the Church but, of the ridicule and derision of his fellow scholars. Not much changes!'

'So, Ollie, I wonder if the judges there said, "Your heliocentric theory *may be true,* Mr Galileo, but your conservative scientific opponents say it's just not *science!*"'

'I wouldn't be surprised. Another strange facet of the Dover case, Ali, is the judge's comment that modern intelligent design theory "violates the centuries-old ground rules of science by invoking and permitting *supernatural causation"*.

'You mean it violates the materialist hypothesis and belief, that controls so much of science and biology today, that *all* Natural phenomena must be explained in terms of purely unintelligent and random or 'ordinary' Natural as opposed to 'super' or higher Natural processes? That scientists are never to be allowed to rationally infer *intelligent* causation in and of the Living World?'

'Yes, exactly. But that view is a *science blocking* and stopping intellectual *dead end.* It is like telling archeologists that they cannot infer intelligent causation, when examining some ancient artifact, because its intelligent causes are knowhere to be seen but can only be rationally inferred. Look, for Plato, Aristotle, Bacon, Copernicus, Galileo, Kepler, Boyle, Newton, Leibniz, Faraday, Maxwell, Planck and many others, most of the greatest *scientists* who have ever lived, there was *nothing* in existence that was *not,* ultimately, intelligent, purposeful and 'super' Natural or spiritual in origin, *not one thing.* After all why should anything at all exist – even just one quark? What could be more 'super' Natural or clever than that? Something at all rather than nothing at all. Those titans of classical thought didn't fill their knowledge–gaps with words like 'mindless', 'unintelligent', 'random' and 'accident'.

'The founding giants of science studied Nature *because* they believed it was *intelligent* and multi-dimensional or spiritual in origin. It was, in their view, its *very intelligent* or super Natural sources and causes that made it both so *intelligent and* scientifically and mathematically *intelligible* and often so elegantly, $E = mc^2$.

'The judge in the Dover case also said that intelligent design or ID is based on "flawed and illogical" arguments. This, again, is peculiar. Is the digital DNA software CODING, that runs all the Life Forms – that is, as Bill Gates put it, "like a computer program but far, far more advanced than any software we have ever created" – an 'unintelligent' phenomenon? Is it a "flawed and illogical" argument to argue that it is not and, based on all we know about code and software construction, that, rationally, it can only have intelligent origins? Are COBOL and BASIC, much simpler than the DNA Coding, and Music notation and

Morse code unintelligent and random phenomena? Is it illogical to argue that DNA CODE like all codes can only have intelligent sources or be intelligently designed? This is just a fact that is governed by probability theory, mathematics and statistics. As Werner Gitt says:

> 'The basic flaw of all [materialistic] evolutionary views is the origin of the information in living beings. It has never been shown that a coding system and semantic information could originate by itself ... The *information theorems predict that this will never be possible.*'[137]

'Are tiny, individual Living Cells, here from the very beginning, that make our spaceships look like wooden spoons 'unintelligent' and accidental phenomena? Has Darwinian biology shown that they are? Is that a rational conclusion to draw when studying them? Or do a powerful, minority in modern science and biology simply claim that they are the stunningly clever expressions of materialism's knowledge-gap-filling pseudo-uber-gods of Mindlessness, Unintelligence and an infinity of 'Accidents' because of a passionate anti–meaning-and-intelligence-in-Nature bias based on naively simplistic, and therefore, unreasonable notions of intelligent causation that looks in their minds like Father Christmas and is therefore, as we'd all agree, absurd?

'Is it an illogical argument to point out that certain features of the Living World are irreducibly complex and could not possibly have arisen in some 'accidental' incremental step by incremental step Darwinian way such as the irreducibly complex blood-clotting cascade, that Michael Behe discussed in *Darwin's Black Box*? Is it a flawed argument to point out that Living Nature's countless Bio-logical (Life-logical) metamorphoses cannot possibly be driven, *right now,* second by second, 24/7, by Life-Forceless, 'dead', zombie 'accidental chemistry' as modern methodologically materialist bio-logy so unreasonably insists because someone artificially synthesized urea in 1828 even though the results of the Life and Soul Forces organ-izing and in-form-ing activities stare us so obviously in the face?

'Is it a flawed argument to point out, as Darwin's peer Alfred Russell Wallace eventually concluded, that the astonishing metamorphosis of the perfectly viable Living Being, the Caterpillar, (with no rational Darwinian style reason to morph into a chemical soup and back into an amazing Butterfly), is not possibly, let alone rationally, explainable in terms of accidental chemistry or 'natural selection'? What possible favorable intermediate stages would there be, that could be 'naturally

---

[137] Gitt, W. 2006. *In the Beginning Was Information.* Master Books, p. 124; emphasis and text in square brackets added.

selected' in such an all or nothing process? Where's the intellectual justification for such claims? Wallace concluded there were none, this is why he eventually abandoned the purely materialistic Darwin–Wallace hypothesis of blind, Life Forceless and Soulless random evolution as being incapable of explaining Life's unfoldment on Earth.[138]

'The judge, in Dover, also held that the modern intelligent design theorists' ideas "have been refuted by the scientific community". This, Ali, is as good as the Court that said to Galileo, "Your conservative, old-school, scientific–consensus opponents, who are implacably opposed to your heliocentric ideas, Mr Galileo, because they threaten the prevailing geocentric paradigm, have *refuted* your arguments because *they say* that the Sun orbits the Earth, and, therefore, that *your arguments are unscientific.*" History really does repeat even if it does not rhyme. Today it's the methodologically materialist, unintelligent design paradigm that is so reluctant to change its mind and to look again.

'Yet, this ignores what Darwin himself said: Firstly that the fossil records did not bear out his theories and that this was: "the *most obvious and gravest objection which can be urged against my theory.*" And, secondly that he was well aware that:

> 'there is scarcely a single point discussed in this volume on which facts cannot be adduced, often apparently leading to conclusions *directly opposite* to those at which I have arrived.'[139]

'Even though Darwin used the word 'apparently' he was, clearly, admitting the possibility that the data could be interpreted in more than one way. The evidence from the discontinuous or new-car-model-like fossil records is far more supportive of the classical view that Natural Selection *confirms* the Species type and does not deviate it away into new forms – for which latter Darwinian process the Fossils have never provided any genuine evidence.

> 'In 1990, Ohio State University biologist Tim Berra published a book intended to refute critics of Darwinian evolution. To illustrate how the fossil record provides evidence for Darwin's theory of descent with modification, Berra used pictures of various models of Corvette automobiles. "If you compare a 1953 and the 1954 Corvette, side-by-side," he wrote," then a 1954 and in 1955 model, and so on, the descent with modification is overwhelmingly

---

[138] Flannery, M.A. 2011. *Alfred Russell Wallace: A Life Remembered*. Discovery Institute Press.
[139] Darwin, C. 1859. *On the Origin of Species*. Emphasis added.

obvious." [He's correct in that it is a great illustration of discontinuous, jumping, that is *intelligent* as opposed to random evolution in operation]. ... Corvettes actually prove the opposite of what Berra intended – namely, that a succession of similarities does *not*, in and of itself – provide evidence for biological descent with modification.[140]

As Richard Milton pointed out:

> '*no transitional species showing evolution in progress between two stages has ever been found*. ... the theory of evolution has become *an act of faith* rather than a functioning science.'[141]

'In *Darwin on Trial* Philip Johnson quoted from a lecture given at the American Museum of Natural History, by Collin Patterson, of the British Natural History Museum, in which he said, to the assembled evolutionary experts:

> '"*Can you tell me anything you know about Evolution, any one thing ... that is true?*" I tried that question on the geology staff at the Field Museum of Natural History and the only answer I got was silence. I tried it on the members of the Evolutionary Morphology seminar, in the University of Chicago, a very prestigious body of evolutionists, and all I got there was silence for a long time and eventually one person said, *"I do know one thing – it ought not to be taught in High School."*'[142]

'Yet, ironically, it is taught in high schools all over the world and those who challenge it are considered to be intellectually deviant and disreputable. To argue that science and biology show that Life is meaningless, purposeless and 'accidental' is considered academically respectable and is, supposedly, properly 'scientific' even though *there's not a shred of evidence that it's true.* To argue, on the other hand, that nothing in our experience of the different kinds of causation, accidental versus intelligent, gives us the right to argue that the Living World and the amazingly complex Living Cells at the base of it all originally arose and, just as importantly, arise right now, by 24/7 'accident' is considered to be in intellectual bad taste! We live in strange times, Ali.

---

[140] Wells, J. 2006. *The Politically Incorrect Guide to Darwinism and Intelligent Design*. Regnery Publishing.

[141] Milton, R. 2000. *Shattering the Myths of Darwinism*. Inner Traditions Bear and Company; back cover, emphasis and text in brackets added.

[142] Johnson, P. 2010. *Darwin on Trial*. IVP Books, p. 28. Emphasis added.

'We continued on quietly for a few moments, then Ali said,

'What I don't understand, Ollie, is why it is, according to the court in Dover, *'scientific' to try to prove that the Living World is unintelligently* designed and, fundamentally, an 'accident', as materialist and Darwinian thinkers seek to do, while it's *allegedly not 'scientific'* to show, using *empirical,* scientific and mathematical theories, that Living Nature could not possibly have arisen or be arising right now 'by accident' as non-materialist and philosophical idealist thinkers hypothesize?'

'I agree, Ali, it is very strange. Young people, perfectly capable of making up their own minds, are, the court orders, *not to be allowed to learn*, in school, that there are highly qualified scientists and thinkers, such as Professor Behe or Dr. Stephen Meyer Ph.D who argue that modern *science can show* (not religious texts) that they, their families and friends are not, after all, Darwinian *'accidents'*.

'And we wonder why so many children and young people are often sad and troubled today. Their spirits are affected by the *ideology* of Darwinian materialism. Something tells them it's a lie to claim that Life–Forceless, Soulless unintelligence and 'freak flukes' built and *build, right now,* trillions of astonishingly dynamic operations happening in every human body every second, this very second, the most intelligently complex and pervasively teleological mechanisms known to humanity – the Soul Life Form Beings of the amazing Plants, Animals and People.'

'This is what the gloomy mental-empire of the design-blind, Darwinian Wall of the Mind is doing to impressionable children and young people, all over the world, in the name of the once noble science of Plato, Bacon and Galileo. It is an intellectual betrayal.

'One set of scientists, supporters of the materialist, 'mindlessness-unintelligence-and-accidents-create-everything-clever-in-Nature', Darwinian unintelligent design or UD school, argue that it is *more 'scientific'* to teach children and young people that they, along with the entire Living World, *are second by second, Darwinian 'accidents'.* It has *never* been a proven theory, rather, it is a dogmatic ideology, a quasi–religion, the ideology of materialism, a horrible, de-Souling, de-intelligencing scheme of cosmic unmeaning and 'accident'. Bizarrely, this sometimes court endorsed ideology of a great delusion or collusion is not considered to be 'anti-religion'. Yet, it is not only unreasonable it is also tantamount to the *abolition* of religion and all spirituality which, despite their various faults, do not preach the Soul-destroying and dishonest (or at best deluded) gospel that people are Soulless, Life-Forceless Darwinian *'accidental-chemical-reactions'*.

'Another set of non-materialist thinkers, modern Idealists and theorists of *intelligent* design, argue that the world's young people are not Darwinian 'accidents' *at all*, and cannot possibly be so, and that they can demonstrate this – not with religious or spiritual texts – but scientifically and mathematically – that is beyond the shadow of any reasonable, evidence-based doubt. They, however, are, at times, even *legally barred* from putting forward their view because they are accused of being religiously motivated.

'This is ironic because there is no doubt at all that some, at least, of their materialist, unintelligent design favoring opponents are motivated by a passionate antipathy towards all religion and spirituality, all purpose and intelligence in Nature, which, in dismissive disregard of all rational inferences and logical deductions to the contrary, and of collective humanity's intuitions and spiritual experiences across millennia as to Life's deeper meanings and higher purposes, they regard as so much make believe and wishful thinking.

'In the Dover case, the judge said that intelligent design *"may be true"*, but "it is not science" because it implicates the 'super' Natural in it's explanations. Yet, there is nothing more 'super' Natural than very Existence itself. Why does anything at all exist? Even one photon? Isn't that pretty 'extra-ordinary', pretty 'super' Natural? Does that mean, then, that we should stop doing science altogether because, in the end, *all explanations* inescapably come back to the Final Source and Uncaused Cause of All things, supreme intelligent Beingness, as Aristotle pointed out more than two thousand years ago – and he wasn't basing his conclusions on the world's religious texts but on Logic and Reason.

'Yet, every day the young are taught in school science and biology class, that they, their families and friends, the entire Living World in fact, are Soulless, Vital-Forceless, Darwinian 'accidents'. This affects their Souls and spirits. It is *not* harmless. It caused me, and many others I know to become materialist–atheists. Hardly an inconsequential decision in one's life.'

'Yet, what's absurd about this, Ollie, is that a concept of an *intelligent* evolution doesn't support anyone's particular religious claims.'

'You're right, of course it doesn't. It simply gives young people the the academically respectful chance to know that some scientists argue that *science* (not religion) can now empirically *show* that the Living World *cannot be rationally explained* or explained-away as a long-running 'accident' – while the young people can learn, at the same time, that the Darwinian, unintelligent design school believes that the

Living World *can be* explained as a never-ending, 24/7, 7/52 *'chemical accident.'* or ACR. Let the young people make up their own minds.

'The bizarre thing is, Ali, that if you put forward falsifiable, *testable scientific hypotheses* that the Living World simply cannot be a random, chance affair, in some jurisdictions this is considered to be 'religion' – but if you teach egregiously non-evidence-based and anti-empirical *unintelligent design theories* of Life and the Living World, like the Darwinian one, that stubbornly disregard the possibility of their falsification, and that imply, for many people, that Life is an unintelligent phenomenon, an 'accident', although you are effectively teaching the *abolition* of the religious and the spiritual, of all that gives Life deeper and higher meaning, you are free to go ahead.

'This is the back to front, Emperor's New Clothes intellectual world we now live in, where, in gloomy, topsy turvy, Darwinia 'accidents', it is insisted, are 'creative' and evolutionary, 'dumbness leads to intelligence', 'mindlessness leads to minds', where the classical, non-materialist or Idealist, 'mind-before-matter-and-mechanism' philosophy and science of Plato, Aristotle, Bacon, Galileo, Newton, etc has been turned all inside out, back to front and upside down.

'However, to question the materialist belief or faith position that the Living World is an ongoing, moment by moment, 24/7 *'accidental chemical reaction',* that has been running for no rhyme, intelligence, purpose or reason, for billions of years, is often considered today to be in as much *bad scientific taste* as Galileo's controversial suggestion that the Sun did not, after all, orbit the Earth was considered in his day. The materialist theory of random evolution by *'accidental-chemical-reaction'* is the bizarre ruling dogma, upheld by the Courts and mandated to be taught in Schools and Universities. Those who question it, like the modern Idealist, Galileo–type heroes of intelligent design in the USA, UK and elsewhere, are mocked or ignored even though their thinking is directly in the classical *intelligent design* tradition of the founding giants of western science – Aristotle, Bacon, Galileo, Kepler, Newton, Boyle, Leibniz, etc.

'Those thinkers were all Idealists and ID Theorists. They were *none of them* materialists. They all believed that relatively intelligent mechanisms like clocks or scientific instruments or extremely intelligent ones like Nature's Laws, the smart-materials-chemicals or Cells or Plants or Animals or People are not 'accidents'. How could they think otherwise? They were steeped in two thousand years of Western rationalism, and it was obvious to them that *intelligent* mechanisms *never* arise from *unintelligent* or 'accidental' sources.

'No medieval or modern clock was ever made by randomly throwing the parts around and no Living Cells, "more complex than Jumbo Jets", have ever been made by sloshing-around pond-water being struck by Darwinian lightning – a fact now proven empirically by, among others, Dembski, Behe and Myer's research.[143] Yet teach children and young people that Life is a random chemical fluke and that wonderful, magical and amazing Soul Beings like Wolves and Eagles, Bison and Elk, Mountain Cats and Bears, Butterflies and Honey Bees, Willows and Oak Trees, and People are all nothing more than 'accidental' genetic replication 'machines' whose sole and meaningless pseudo 'purpose' is simply to replicate in order to replicate for no reason at all except to replicate and few eye brows are raised.

'Few seem to notice that all genes are made of, fundamentally, is complex compounds of the already extremely plentiful atomic elements of Oxygen, Carbon, Hydrogen and Nitrogen which have no need or inherent tendency to replicate themselves – selfishly or unselfishly. They are quite happy as they are. No brick ever 'accidentally' caused a house to arise. No nut or bolt ever 'accidentally' became a car or airplane and no amino acid has ever 'accidentally' caused a Life–Form to arise. There is something more, call it 'intelligence', Life–Forces, Soul–Forces, call it whatever we like, than bricks that causes houses to arise and something more than amino acid 'bricks' that causes Life–Forms of stunningly intelligent, Living, Breathing, Self-Reproducing complexity to be. It is quite clear that outside topsy-turvy Darwinia 'machines' or 'robots' or 'computers' or CODEs and software to run them *never arise* 'by accident'.

'Yet, as long as you do it diplomatically, if you explain to children and young people, in biology class today, that they and the entire Living World are, according to modern materialist and Darwinian theories, at base, ongoing, second by second zombie 'accidents' without superordinate Souls or Life–Forces (due to the synthesis of urea in 1828) or meaning or purpose of any kind, you'll probably be ok, perhaps even applauded. Find a tactful way to tell the children and young people that:

> 'You, your joys and your sorrows, your memories and your ambitions, your sense of personal identity and free will, are in fact

---

[143] Dembski, W. 1998. *The Design Inference.* Cambridge University Press. Behe, M. 1996. *Darwin's Black Box.* Free Press. 2007. *The Edge of Evolution: The Search for the Limits of Darwinism.* Free Press. Meyer, S. 2009. *Signature in the Cell: DNA and the Evidence for Intelligent Design.* Harper One.

no more than the behavior of a vast [accidental and meaningless] assembly of nerve cells and their associated molecules. As Lewis Carroll's Alice might have phrased it: "You're nothing but a pack of neurons."'[144]

and you'll probably get away with it. Such explanations have caused a twelve year old I know to conclude that 'God or some meaningful, higher Power doesn't exist and that Life is meaningless, unintelligent and accidental'. He's only twelve and he's now a 'materialist skeptic'.

'Suggest, on the other hand, that some scientists are now arguing that classical science, mathematics and information theory can show that *it is impossible to logically* explain the second by second complexities of the Living World as 'accidents' and that the signs of *intelligence and purpose* are pervasive in Nature and you may be charged with teaching 'religion'.

'You see Ali, there is a fight on for the very Soul of science. For some, the word 'science' now means materialism. It is a dark, de-Souling philosophy for it is in a struggle with the supernatural and the spiritual. It seeks to cut humanity off from its intelligent spiritual sources, preaching that they don't exist, that only matter exists and only matter matters. Conceptually, it chops Nature in half, de-intelligences it, de-purposes it, de-Souls it, de-ments it, de-hallows it and hollows it out, throwing away all that is 'super', intelligent, magical and sacred, keeping only the so-called 'ordinary' Natural chemical parts and, in the process, turns it into a de-Souled, hollow, wax–work, zombie mockery of what it truly is.

'For others, though, with more open minds, science consists of many related and unrelated ways of exploring *all levels* and dimensions of Holistic Reality – so-called 'ordinary' Nature and 'super', higher Nature, Consciousness and its Expressions, Intelligence and Chemicals, Life and Form, Soul and Mechanism, Ghost and Machine, Subject and Object – a two sided Unity – inner and outer at all levels and dimensions.

'True, open-minded, classical science doesn't reflexively dismiss vast sweeps of human experience (psychic, religious, mystical and spiritual experiences, out of body experiences, near death experiences, memories of previous existences, energetic and spiritual healing) because they didn't take place under laboratory conditions

---

[144] Crick, F. 1995. *The Astonishing Hypothesis*. Scribner; p. 3.

and then with an astonishing lack of integrity claim that 'science' has 'proved' that the Living World is a chemical 'accident', not because of any genuine scientific evidence, let alone laboratory evidence, but simply because the *intelligently* causal dimensions, of nearly universal human intuition, classical rational inference and sometimes direct psychic, mystical and spiritual experience and revelation, not 'belief', cannot be physically seen in microscopes or telescopes.*This does not constitute laboratory evidence* for the truth of materialist, 'life-is-an-accident' evolution.

'There is *no laboratory evidence* for Darwinian-style random evolution. This is one of the things that gets me down about materialism Ali – it *lacks intellectual integrity.* If the sciences of matter and energy (physics, chemistry, biology) are *only* about questions that can be 'tested' and 'proved' in laboratories, then we need to list all those subjects that can be treated in this way.

'The materialist dogma that *'Life-is-a-purposeless-24/7-accidental-chemical-reaction'* isn't one of them. Darwinian evolution 'by accident' should, correctly, be labeled scientific *speculation*, not 'proven' science. There is no evidence at all that *random* genetic mutations are the sources of the clever unfoldment of the Living World, across millions of years, that the amazing Fossils make visible. The Darwinian evolution by 'genetic-accident' hypothesis has become a rigid *belief and a dogma.* There is no 'proof' for it other than the materialist conviction that (a) there *are* no possible intelligent causes of the Living World (*atheistic* Darwinism); or alternatively (b) that even if there *were*, science would not be able to detect them (*theistic* Darwinism).

'This *doesn't make sense.* If science is capable of detecting *unintelligent* causation or 'unintelligent design' in Living Nature it is as well equipped to detect *intelligent* causation, 'intelligent design' or 'intelligent evolution', across vast time. This is precisely what today's Idealist, intelligent design theorists, thinkers like Behe, Dembski and Meyer, among others, have successfully demonstrated.

'But an ideologically–materialist science with, as Lewontin puts it:

'a prior commitment, a commitment **to materialism.** ... [A] **materialism [that] is absolute**, for we cannot allow a Divine foot, [or *intelligent* causation], in the door.'[145]

---

[145] Richard Lewontin (1997) Billions and billions of demons (review of *The Demon-Haunted World: Science as a Candle in the Dark* by Carl Sagan, 1997). *The New York Review*, January 9, p. 31. Emphasis and text in square brackets added.

'has to insist that Nature, in its totality, is 'unintelligent' and accidental doesn't it, Oliver?'

'Yes, sadly, you're right, it does, but remember, as Chris Carter says:

> 'the difference between science and ideology is not that they are based on different dogmas; rather, it is that scientific beliefs are ***not*** held as dogmas, but ***are open to testing and hence possible rejection.*** Science cannot be an objective process of discovery if it is wedded to ***a metaphysical belief that is accepted without question*** and that leads to the exclusion of certain lines of evidence on the grounds that these lines of evidence contradict the metaphysical belief.'[146]

'The problem is that methodological–materialism in science has become a science blocker, a quasi religion, an ideology. An intellectual stance that makes science dishonest or at best delusional. A dogma that makes words like 'intelligent' and 'purposeful' and 'teleology' *meaningless* – because if Living Nature's breathtakingly dynamic activity, which makes the clever and *intelligent* functioning of Jumbo Jets and Space Shuttles look like the functioning of babies rattles, is not 'intelligent' and 'purposeful' or 'teleological', then nothing is, and we can no longer have any kind of meaningful or intelligent conversation because these important words have become meaningless.

'For now, the materialist, Darwinian theory of meaningless, and Soulless evolution by 'accident' is still the 'anti-Galilean-consensus' view. It has even convinced a court (Dover, 2005) into ruling that to detect *unintelligence* and *purposelessness* in Nature using Darwinian-style *unintelligent design theory* is 'scientific' but to use intelligent design theory to detect *intelligence and purpose* in Nature is not.

'This is an unfortunate result. It is now official legal policy, in that school district, that children are only allowed to hear of the *Darwinian* concept of evolution, that the magical and mysterious unfoldment of the Living World, over vast time, that the Fossils make visible, is a wholly 'unintelligent' and random process. They are not allowed to hear that other physical-scientists and philosophers argue, in line with Darwin's own advice –

> 'that there is scarcely a single point discussed [in *On the Origin of Species*]… on which facts cannot be adduced, often apparently

---

[146] Carter, C. 2010. *Science and the Near Death Experience.* Inner Traditions; p.237, emphasis added.

leading to conclusions *directly opposite* to those at which I [Darwin] have arrived.'[147]

'Yes, probably Darwin didn't entirely mean what he said because he qualified it with the word 'apparently' but he was acknowledging the possibility that his theory might not be correct. The Fossils, certainly, have never confirmed it. In fact, as Darwin himself said, this was the 'most obvious and gravest objection which can be urged against my theory.'

'Yet, even now, in many schools, children and young people have to continue to leave their minds and their commonsense at the door of biology class while they absorb the Soul corroding materialist claim that they are 'accidents'. Like that innocent, bright, intelligent twelve-year-old I know, who now believes he's a Darwinian 'accident' because that's what his teachers have told him. He's already a materialist–atheist skeptic. This is sad.

'Of course, in other settings my young friend and his fellow classmates will be taught about higher things, about religion and spirituality, about the good, the beautiful and the true – and… *they probably won't believe it* because physical–science is regarded by so many in our culture as the highest source of 'real knowledge' and if the materialist ideology that has captured western science and biology and causes it to say absurd and unreasonable things, like: "Minds are accidents made of atoms", "unintelligence leads to intelligence," "accidents build the cleverest 'machinery' known to humanity," "the digital, computer-type DNA CODING that helps to run all Living things is the [blazingly intelligent] result of utterly unintelligent and *'accidental'* chemistry and the Living World is literally an ongoing, 24/7 *"accidental chemical reaction"* – then, absurd and false as these ideas are, children and young people, innocently trusting their teachers, and barred, by order of the Courts from learning of any alternative and *counter-balancing* scientific views, will believe these untruths.'

We walked on, by the river, through the meadows.

---

[147] Darwin, C. 1859. *On the Origin of Species*.

10 Head to Head: Unintelligent Design Versus Intelligent Design

DNA CODING is 'like a computer program but far, far more advanced than any software ever created' – B. Gates. Is the enigma of the DNA CODED digital information in the Cell better explained by materialist 'accidents-create-codes' Unintelligent Design theories or by idealist 'intelligence-creates-codes' Intelligent Design thinking?

'Darwin mentioned his famous 'Life-is-a-chemical-accident' idea to his friend, Joseph Hooker, saying that perhaps, "in some warm little pond" *chemical accidents* might produce *Living Cells – the most complex micro-miniaturized systems known to humanity*.

'This idea, however, like all materialist thought since, overlooked the fact that there is *nothing* in the laws of *'chance' chemistry* to cause us to believe that these can ever 'create' or 'evolve' anything other than random rocky screed showing *complexity* but lacking *specificity* or crystals showing *specificity* but lacking in *complexity*.

'Mechanisms like cars, TVs and computers are (unlike crystals) *highly complex* and (unlike random, rubbly screed) *highly specific*. Cars, TVs and computers are *specifically complex* and rely on *specific* assembly, running and maintenance information or they will not work. Such information *never* arises by random, chance processes. It is always provided in intelligently designed instruction manuals and computer codes that run robotized machine tools in the case of the manufacture of cars, etc, and by the DNA digital CODE in the case of the manufacture of the Life Forms. Jumble the coding software or the in-form-ational instructions and the mechanisms won't work – attach the wheels of a car to the roof, and it's useless.

'As Dr. Stephen Meyer says, "Our *uniform experience* affirms that *specified* information [as opposed to random, *unspecified* information] – whether inscribed as hieroglyphics, written in a book, encoded in a radio signal, or produced in a simulation experiment – *always* arises from an *intelligent* source, from a mind and not a strictly material process".[148]

'As Meyer explains, our universal experience teaches us that intelligently and *specifically complex* information, of the type that runs the computers on Jumbo Jets or that runs all Living Cells, "more complex

---

[148] Meyer, S. 2009. *Signature in the Cell*. Harper One; p.347, emphasis and text in square brackets added.

than Jumbo Jets", via the DNA CODING, *never* arises by accident but is *always* intelligent in origin. Random, *unspecific* complexity describes rocky rubbly screed at the foot of a mountain or pretty wave marked patterns on sandy beaches, but never intelligent mechanisms. As Leslie Orgel said:

> 'Living things are distinguished by their ***specified complexity.*** Crystals such as granite fail to qualify as living because they lack complexity [being specific but not complex]; mixtures of random polymers fail to qualify because they lack specificity [being highly complex but randomly so].'[149]

'Specified simplicity or specified *un-complexity* describes mechanisms like platted ropes or crystals where a simple and specific pattern repeats, according to a simple algorithm, like ABC, ABC, ABC. This kind of *specified un-complexity* never builds or gives expression to complex mechanisms. Cars, Computers, TVs, Jumbo Jets and Life Forms are *never* built out of such simple, if specific, patterns or algorithms.

'The digital CODING of the genomes, written in deoxyribonucleic acid ink, is not a simple and repetitive instruction like 'ABC + ABC + ABC' (which can be written very simply as 'ABC repeats'), but conveys the massive amounts of genetic information, sentence after sentence after sentence for translation by the ribosomes at *super-computer speeds* into Living Forms: TTAAGGCC ATGCAGGA CAACGGTT CCTTGTAC CAACGGTT TATAGGCC AGGAATGC TATAGGCC CCTTGTAC. This is nothing at all like the simple algorithm or instruction: 'ABC repeats'. As even Dawkins says of genetics, without, accepting the intelligent design implications:

> 'It is ***pure information***. It's ***digital information***. It's precisely the kind of information that can be translated digit for digit, byte for byte, into any other kind of information and then translated back again.'[150]

'And this is precisely the kind of *specific,* non-random information that never arises 'by accident'. The laws of probability and statistics

---

[149] Orgel, L. 1973. *The Origins of Life.* John Wiley; p. 189. Emphasis and text in square brackets added.

[150] R. Dawkins, 2008. *Life: A Gene-Centric View, Craig Venter & Richard Dawkins: A Conversation in Munich. The Edge.*

forbid it. This is made clear in W. Dambksi's book *No Free Lunch: Why specified complexity cannot be purchased without Intelligence* (2007)The intelligent, DNA CODED instructions in the genomes of the Plants, Animals and People are to Jumbo Jet assembly instructions as Jumbo Jet assembly instructions are to the assembly instructions for simple wooden toys. It is no longer intellectually credible to argue that such *intelligent* complexity is the product of 'accidental chemistry'.

'Entropic accidents don't build *intelligently* complex mechanisms of any kind. They erode, corrode, dissolve, grind down and degrade. The magical and Soulful Plants and Animals in these fields, built by the ribosomes, run by the clever DNA CODING, "more advanced than any software ever created", are not DNA CODED *'accidental chemical reactions'*.'

'How can modern biology think in this way, Ollie? How can our thinking have become so unreasonable?'

'Look, it is clear today that the materialist hypotheses of Consciousness, Life, Mind and Living Nature as a whole stand refuted – they do not work. The exciting news is that things are changing, remember how, encouragingly, the judge, in the Dover case, held that intelligent design *"may be true"*.'

'Hold on, Ollie, the judge in that case actually endorsed the Darwinian, *unintelligent* design ('accidents-build-machinery-and codes') view as being 'scientific', while holding that classical Idealist, ('intelligence-builds-clever-mechanisms') *intelligent* design thinking was not 'scientific'.'

'Okay, true, but, firstly, he was wrong because most of the greatest scientists that have ever lived were idealist, intelligent design thinkers from Pythagoras to Max Planck via Bacon, Galileo, Newton and almost everyone in between. Secondly, it seems that he confused his personal opinion as to the scientific merits of intelligent design with the issue as to whether it is 'science' at all. Many have argued, from the very first, that Darwinism itself is *not 'good science',* that they are not persuaded by its arguments which are not borne out by the Fossil Records[151] – with Darwin himself admitting that the lack of Fossil evidence was the "most obvious and gravest objection" to his theory – but they never said that it was not 'science' at all, even if some, famously Karl Popper, said that the Darwinian theory was pseudo-science because of its slippery refusal to allow for the possibility that it could be falsified or

---

[151] See, for example, Denton, M. 1985. *Evolution: A Theory in Crisis*. Burnett Books. Wells, J. 2002. *Icons of Evolution*. Regnery Publishing. Latham, A. 2005. *The Naked Emperor: Darwinism Exposed*. Janus.

refuted, which arguably has already occurred on numerous occasions and in numerous ways.

'Really, Ali, this is a dispute in which some scientists argue that only an infinity of accidents and unintelligent matter exist and that it is the job of science to *intelligently* explain how Nature's mechanisms all unintelligently and purposelessly function, second by second, 24/7 – as a result of entirely mindless and accidental processes – the materialist view. Other scientists hypothesize that total Reality is intelligently multi-dimensional and consists not only of matter but also of *consciousness and purposeful intelligence* and that science can explain for us how Nature's incredibly clever mechanisms *work*, how they are put together and what they are made of (encompassing two of Aristotle's famous four causes – those of materials and mechanisms), without, necessarily, yet being able to explain *how or why* they arose in the first place and continue to do so, right now, clever second by clever second, 24/7. That there is something more to Life than 'mere physics and chemistry' – the Idealist view (and touching on Aristotle's last two causes, formal and final – that is ideas and purposes).'

'The Idealist, intelligent design thinkers argue that one of the basic tasks of science is to carefully discriminate between those features of the Natural and Living Worlds which are random and *accidental* and those which are purposeful and *intelligent*. The materialist reply to this is that there are *no* features of the Natural and Living Worlds which are intelligent and purposeful, let alone even truly conscious, because the whole thing is one big 'unintelligent, ghostless, 'accidental', unconscious, zombie machine', from quarks to galaxies, all the way up and all the way down.

'The idealists reply that it is humanity's universal experience that 'machines' and 'machinery' simple or complex never arise 'by accident'. The materialists, however, are *not interested in discussing the evidence* that this is so.

'What is a shame though, Ali, is that *both* UD and ID thinking have their places in science. Unintelligent design thinking is great for understanding the random features of the Natural world, such as randomly evolved river valleys and other features of the landscape.

'On the other hand intelligent design thinking is great for helping to make sense of Nature's intelligent features like DNA CODING, 'more advanced than any software', Living Cells, 'the most complex systems known to man', Cows and Sheep, Frogs and Buttercups, Honey bees, Apple Trees and People none of which can reasonably be thought of as unintelligent or accidental effects of unintelligent and accidental causes.

'In his seminal book, *Signature in the Cell: DNA and the Evidence for Intelligent Design*, Stephen Meyer Ph.D. explains that the idea that intelligent design or ID is 'not science' is a *myth*. This is true. It is a myth that is, not surprisingly, promoted by its ideologically–materialist 'unintelligent design' opponents.

'He says:

> 'As I present the evidence for intelligent design, critics do not typically try to dispute my specific empirical claims. They do not dispute that DNA contains *specified information,* or that this type of information *always* comes from a mind, or that competing materialistic [unintelligent design] theories have failed to account for the DNA enigma. Nor do they even dispute my characterization of the historical scientific method ... in formulating my case for intelligent design as the best explanation [versus "accident"] for the evidence [the digital DNA CODING]. Instead ... Critics simply insist that intelligent design "is just not science".[152]

'Meyer says the objection that 'ID isn't science' is really code for: *'It isn't true, it's disreputable,* and *there is no evidence for it'.* He says that because the claim that intelligent design is 'not science' has been repeated so often and with such presumed authority that many people have rejected it before even considering the arguments for it and the evidence in it's favor. He explains that he eventually realized that in order to make his case for intelligent design in the Living World he would have to defend the theory against the charge that it's 'not science', 'it's disreputable' etc.

'Meyer does this by explaining that the holdall word science encompasses many different disciplines and that it is not intellectually correct to say that modern Intelligent Design or ID theory is 'not science' any more than it would be correct to claim that the Darwinian hypothesis of Unintelligent Design or UD is 'not science'. He explains that some sciences put their theories to the test by making predictions, others by making comparisons between the predictive or explanatory capabilities of competing hypotheses. Some scientific fields make speculations that currently cannot be tested at all. Some sciences study only what can be seen, others make inferences about entities that cannot be seen. 'Some sciences reason deductively; some inductively; some abductively ... Some use the method of multiple competing hypotheses.'[153]

---

[152] Meyer, S. 2009. *Signature in the Cell*. Harper One; p. 396. Text in square brackets and emphasis added.
[153] Ibid., p. 401.

'Meyer explains that the case for intelligent design, like other scientific theories, is based on empirical evidence. It's not a 'religious' theory and, contrary to the claims of one of the expert witnesses in the Dover trial, R. Pennock, ID theorists have developed solid and specific empirical arguments to back up their theories. In *Signature in the Cell: DNA and the Evidence for Intelligent Design* Meyer puts forward an argument for intelligent design based on the intriguing modern discovery of ***digitally coded information*** in the Cell. Although it is a decisive piece of evidence it is not the only evidence in favor of intelligent design in Nature.

'Meyer notes that other scientists believe that the *irreducible complexity* of molecular machinery and circuitry in the Cell points to intelligent design. Some see the *information processing system* of Cells as evidence of intelligent design. The fine tuning of earth's systems and their mysteriously self-regulating nature, that James Lovelock pointed out, in the Gaia Hypothesis, all indicate intelligent, not mindless and accidental, processes. Others regard the stunningly precise fine tuning of the laws and constants of physics as evidence of intelligent design. Meyer says:

> 'Critics may disagree with the conclusions of these design arguments, but they cannot reasonably deny that they are based upon commonly accepted observations of the natural world. Since the *term "science" commonly connotes* an activity in which theories are developed to *explain observations of the natural world,* the empirical, observational basis of the theory of intelligent design provides a good reason for regarding intelligent design as a scientific theory.'[154]

'Critics of ID have said, Ali, that modern intelligent design theories, as they relate to the Living World, are not testable, by way of claiming that they are not scientific in comparison to their rival Darwinian and other materialist theories of Unintelligent Design.

'Meyer, however, says:

> 'Contrary to the repeated claims of its detractors, the theory of intelligent design *is* testable". He continues, it "is testable by comparing its explanatory power to that of competing theories. Darwin used [precisely] this method of testing in *On the Origin of Species.*'[155]

'This method, of *comparing the explanatory power* of competing theories is fundamental to Darwinian thinking which always drew its

---
[154] Ibid., p. 403, emphasis added.
[155] Ibid., p. 404.

greatest strength from comparing its explanation of the Living World with the explanation of the origins of Living Nature provided by misplaced early modern attempts to treat the Bible creation story in Genesis as though it might be a literal description of creation in a physical–scientific sense, as opposed to being *literally true* in a religious or spiritual sense. The creation story of the Rolls-Royce that goes: "Rolls and Royce said, *Let there be magnificent Rolls-Royce Cars!* And lo there were," while true in a spiritual or poetic sense is not usefully true in a technical sense. Equally, the world's spiritual creation stories are simply non-technical descriptions of the fact that the Living World's sources and evolutionary origins are purposeful and intelligent, not mindless and accidental.'

'So, Meyer is saying that we have to decide whether the explanatory power of the materialist belief that random events *unintelligently* created the DNA GENOMIC CODING is more scientifically reasonable (or even remotely possible) or whether the modern idealist hypothesis that the DNA CODING and the digital information in the Genomes is a manifestation of *intelligent* laws, forces and agencies is more scientifically reasonable. You see we have only two choices:

(1) 'Accidents' created the DNA CODING, or

(2) 'Intelligence' created the DNA CODING.

'At the core of this argument about unintelligent versus intelligent design, random versus intelligent evolution is the dispute between the materialists and the idealists as to whether total Reality is mono-dimensionally mindless and accidental or multi-dimensionally intelligent and purposeful. Some Darwinians claim that the DNA CODING sequences simply arise from the operation of 'ordinary' Natural chemical laws, not of 'super' or Higher Natural laws, forces and agencies.

'The idealist, intelligent design thinkers point out, however, that this is no better than claiming that Shakespeare's words arise from the chemical laws of the ink and the paper he used – as opposed to coming from his mind and intelligent intentions. As Meyer explains, in *Signature in the Cell: DNA and the Evidence for Intelligent Design,* the sequence of the DNA letters A, C, T and G on a single strand of DNA is **not determined by any chemical laws** any more than Shakespeare's words and sentences were determined by the composition of the paper and ink with which he wrote. It is only *because,* on a single strand of DNA, the DNA letters can be in whatever coding order is required to convey the relevant instructions that complex coding information can be written with them.

'Chemist, Michael Polanyi pointed out that *if the precise sequences* of the letters of the DNA CODING were *physically* determined, that is simply by the bonding laws of chemistry, "then such a DNA molecule would have no information content ... [The sequence of the DNA letters] *must be as physically indeterminate* as the sequence of words is on a printed page."[156]

'Materialist biology refuses to pay any attention to the intellectual and obvious intelligent design implications of this point, *does not wish to discuss the evidence* for intelligent design, and simply says: 'We can't see any intelligent designers anywhere.' To which the idealist, intelligent design thinking biologists reply:

> 'We can't see any either. However, based on what we know of digital codes and software which (a) never arise 'by accident' and (b) *can* never arise 'by accident' we are prepared to rationally infer and intuit their existence whereas you are not.'

'The materialist thinking scientists don't like this because they feel that admitting the possibility of intelligent causation in Nature is but one step away from admitting that the Cosmos might be meaningful and intelligently multi-dimensional after all, a conclusion that they seem to find depressing rather than stimulating and exciting. Some even seem to feel that this would be but a step away from direct religious rule. This, however, is absurd because admitting that the Living World or the DNA CODES are examples of intelligent design or intelligent evolutionary processes at work does not confirm a particular religious or moral view of the world. It just tells us that existence is intelligent not 'dumb'. We have to look elsewhere for our insights as to how we should live. An intelligently designed or intelligently evolved Nature is still one that is 'red in tooth and claw' so it does not provide a wise basis for ethics, religion or spirituality.'

'So we're back to the Mystery of Existence again, Ollie?'

'True, in the sense that we can't, any of us, say *why* there should be a Cosmic Mystery of Existence at all, *why* there should be something at all rather than nothing at all, not even one electron. Simply saying it's because of a Big Bang doesn't tell us whether such an event is a stupendous 'accident' or an expression of stupendous power and intelligence. Our choices are simply between saying: materialism's gods of 'mindlessness, unintelligence and accidents' created all that is,

---

[156] Michael Polanyi, "Life's Irreducible Structure," *Science* 160 (1968), 1308–12, quoted by Wells, J. 2006. *Darwinism and Intelligent Design*. Regnery Publishing; pp. 100–101.

including the Life Forms and the DNA CODING to help run them, or that idealism's gods of 'purposefulness, meaning and intelligence' created all that is – gods to which collective humanity gives labels such as Supreme Being, God, Brahman, Allah, Buddha Nature, the Tao or simply the Great Mystery.

'I think, Ollie, our own minds and intelligences should be clues enough and the fact that there is no real–world evidence anywhere that 'accidents' ever build anything clever.'

'I agree. Saying that 'accidents' build mechanisms and design digital CODES is as counterfactual as claiming that the world was literally made out of spit and clay in six busy days by a god with a beard. It is quite clear that the real-world 'evolutionary' powers of Life-Forceless, chemical accidents are confined to random and unintelligent effects – river valleys, rocky screed, random drifts of autumn leaves, forest fires and pretty wave marked patterns on sandy beaches. They are *never* known to build *complex machines* or to develop incredibly clever computer–type machine codes to run them.'

'This is why I favor the intelligent or 'super' or 'higher' Natural or Spiritual Mystery of existence solution to the enigma of the sophisticated, digitally CODED information in all Living Cells. A Mystery that all our sciences will continue to explore for hundreds of years to come but with the stimulating knowledge that it's an intelligent Mystery not a mindless and unintelligent one. Meyer explains that in making his scientific, 'intelligence-before-codes' case, in chapters 8 to 16 in his book, he *tested* the theory of intelligent design exactly as Darwin tested his theory – that is by comparing the explanatory power of his theory against that of several other classes of explanation. Meyer does the same by explaining how all the materialistic theories as to how DNA CODING might arise 'by accident' are wanting (in fact, they are intellectually hopeless, although he's too polite to say so) and concludes:

> 'That the theory of intelligent design can explain the origin of biological information [DNA Code] ... better than its materialistic [unintelligent design] competitors shows that it has passed an important scientific test.[157]

'In other words, the scientific theory of intelligent design, when compared with its rival unintelligent design theories provides the better explanation of the DNA digital CODING enigma. It confirms

---

[157] Ibid., p. 404, emphasis added.

that Life's CODING "more advanced than any software we have ever created," has *intelligent* origins whereas its rival, materialistic, unintelligent design theories, based on very firm commitments to purely materialist and 'accidental' cause explanations hypothersize that it is an unintelligently derived and chance phenomenon.

'The problem that today's idealist theorists of intelligent design such as Michael Behe, William Dembski and Stephen Meyer are up against is that so many scientists and non-scientists alike now believe that science means materialism which, by definition, rules out intelligent causation in Nature, *not as a matter of empirical evidence* but as a matter of dogma.

> '*The long-standing confusion of materialism with science* is what largely accounts for the persistent social taboo responsible for the ignorance and dismissal of the **substantial amount of evidence that proves materialism false.**'[158]
>
> 'Science is a methodological process of discovering truths about reality. Insofar as science is an objective process of discovery, it is, and must be, metaphysically neutral. Insofar as science is not metaphysically neutral, but instead weds itself to a particular metaphysical theory, such as materialism, *it cannot be an objective process for discovery.* There is much confusion on this point, because many people equate science with **materialist metaphysics,** and phenomena that fall outside the scope of such metaphysics, [such as the evidence for intelligent design and intelligent evolution] and hence cannot be explained in physical terms, are called 'unscientific'.[159]

'The intellectual consequence of this view is that if intelligent and purposeful causation is a part of reality it is a *priori excluded* as a matter, not of evidence, but of materialist ideology – almost like a dogmatic and fundamentalist religion.' However, as Carter says:

> 'the difference between science and ideology is not that they are based on different dogmas; rather, it is that scientific beliefs are *not* held as dogmas, but *are open to testing and hence possible rejection.* Science cannot be an objective process of discovery if it is wedded to *a metaphysical belief [such as materialism] that is accepted without question* and that leads to the exclusion of

---

[158] Carter, C. 2010. *Science and the Near Death Experience.* Inner Traditions; p. 238.
[159] Grossman, N. 2002. *Who's Afraid of Life After Death?* p.10–12; emphasis and text in square brackets added.

certain lines of evidence on the grounds that these lines of evidence contradict the metaphysical belief.'[160]

'The reality that codes, of whatever kind, simply cannot arise 'by random chance' profoundly contradicts the materialist faith and belief in unintelligent and accidental causation. Modern, methodologically materialist scientific thinking, which functions almost like a kind of pseudo-religion, cannot tolerate the empirical evidence that materialist theories for the origins of the DNA CODING, 'more advanced than any software ever created', don't work and, therefore, stand refuted because, firstly, codes and software systems are never known to arise by accident and, secondly, modern information theory predicts that they cannot do so[161] – it's not merely improbable, it's impossible.'

'So, Ollie, the evidence is that Life's CODE's sources, however weird or spooky that might seem to some, must be intelligent not mindless and accidental.'

'I agree Ali, it might be weird or spooky or exciting and amazing but at least it's consonant with the data of reality. This is why materialism cannot explain all the most interesting aspects of existence.'

'Like?'

'Like consciousness or the universal witnessing awareness for example – the universal Experiencer, Subject and Knower in all Beings. Materialism cannot explain Subjectivity or Sentiency. How can it? In materialism there is no one actually there to sense anything, there is no true Mind, Soul or 'ghost in the machine'. As Dennett says, in materialism: "We're all zombies. Nobody is conscious."[162] Materialism cannot make any sense of Life, Mind, or Soul, nor digital DNA CODING because it insists that they are all 'unintelligent and accidental phenomena'.

'Materialism can only ever provide half–explanations of all the most interesting phenomena of existence. The materialist idea of 'Nature' is a horrible, de-Souled mockery of true, magical, sacred Living Nature. This is because like desert islanders trying to explain an unknown Rolls-Royce as an 'unintelligent' mechanical 'accident,' it's explanations don't work.

'What confuses us, however, is that because methodologically–mate-

---

[160] Carter, C. 2010. *Science and the Near Death Experience*. Inner Traditions; p. 237, emphasis and text in square brackets added.

[161] Gitt, W. 2006. *In the Beginning Was Information*. Master Books, p. 124; emphasis and text in square brackets added.

[162] Dennett D. 1992. *Consciousness Explained*. Back Bay Books, p. 406. Quoted by L. Dossey: 2010. 'Is the Universe Merely A Statistical Accident?' *The Blog. Huffington Post*.

rialist science is so successful at working with Aristotle's first two causes, matter and mechanism ('made of' and 'how things work'), and, as a result of this, providing us with so many brilliant technologies, like harnessing electricity, creating TV, internet, unraveling the genomic codes etc – (like desert islanders cleverly working out what an unknown Rolls-Royce is 'made of' and how it 'works' and how to use it for their own ends) – we begin to think that this brilliantly clever if somewhat faustian science has provided us with a full explanation of total Reality, whereas it has only provided a half–explanation – that of 'matter' and 'mechanism' alone but never one of the true Life, mind, soul and meaning of things.

'Cleverly describing electricity and magnetism and DNA CODING and Living Cells and the Rolls-Royce's workings and materials doesn't tell us anything about their intelligent and purposeful Final Sources and Causes. To describe materials and mechanisms, while clever and useful, is but a half explanation, not a full one. It doesn't tell us why the Rolls Royce (or Big Bang or Life and Mind on Earth) exist or what for. The island's materialist thinkers try to get around this by saying that the magnificent Rolls-Royce (our metaphor for the Living World and the Universe as a whole) doesn't have any formal and final causes, it arose 'by accident'. Methodologically materialist science and biology are based, philosophically, on mindlessness-unintelligence-and-accidents-of-the-gaps thinking. The governing hypothesis of mainstream western science and biology today is that all causation in Nature is 'unintelligent', 'purposeless' (or lacking in teleology) and 'accidental'. This is why the modern non-materialist hypothesis of intelligent causation in and of the Living World is taboo.

'The key point that materialist thinking misses – and which it cannot account for – is that the *intelligent information* in the DNA CODE itself, written in deoxyribonucleic acid or DNA 'ink', is no more 'made of' deoxyribonucleic acid than the clever digital information, meanings and instructions in COBOL and BASIC are 'made of' the paper and *ink* or magnetic media on and with which they are conveyed, or Shakespeare's plays and Mozart's music were or are 'made of' the paper and *ink* on which they were initially written down and subsequently stored.

'Mozart's music manuscripts and Morse Code and COBOL and BASIC and DNA CODE all convey complex and intelligent *meanings* and *instructions – intelligent information* which, our universal experience teaches, only *ever* comes from *intelligent* sources. We can't get at the *intelligent information,* which is mental in origin and nature, and the *meanings and instructions* contained in *any* of these CODES by materialistically analyzing the chemicals or inks in which they are written or the media and means by which they are stored and conveyed. We can only de-code them by understanding the *meanings* and *intelligent information* that they convey, (regardless of which kind of 'ink' they are written in, be it deoxyribonucleic acid or some other kind). *All codes* and instruction sets, experience teaches us, *only ever* come from *intelligent* sources, never mindless or accidental ones. No code was ever formed by randomly throwing 'zeros' and 'ones' in the air or randomly mixing the nucleic acid letters "A", "C", "T" and "G". As Werner Gitt says:

> "It has never been shown that a coding *system* and semantic [intelligent] information could originate by itself… *The information theorems predict that this **will never be possible***. A purely material origin of life is thus [ruled out]"[163]

'That's not: "the world's religions predict this will never be possible" but: "The information theorems predict that this will never be possible". In other words we may not like the conclusion that the digital coding in Living Cells could not arise by 'accident' but it's just a scientific and statistical fact.

'But why would we resist that conclusion, Ollie? It doesn't make sense. Surely we'd be pleased to discover that Nature is, after all, intelligent and purposeful rather than meaningless and 'accidental'?'

'For all sorts of psychological and emotional reasons. Many of us have grown used to the dismal materialist idea that the amazing Living World is one big, unintelligent 'accident', allegedly running second by second for millions of years for no reason or meaning at all. It was not an easy let alone a rational idea for our culture to get used to, over the last one hundred and fifty gloomy Darwinian years, but gradually we did. Now other intellectuals arrive on the scene and say:

> 'Ladies and gentlemen, mainstream science has claimed for generations now that you are Soulless, Vital–Forceless, 'accidents' of chemistry. However, some modern, cutting edge scientists have now realized that this is simply not possible. The Living World is not a Soulless and Life–Forceless 'accident' after all.
>
> 'Not only are human *minds* and *intelligences* a clue that that's unrealistic, but we now know that many biological mechanisms are *irreducibly complex*.[164] This means they cannot arise in the blind

---

[163] Gitt, W. 2006. *In the Beginning Was Information: A Scientist Explains the Incredible Design in Nature.* Master Books, p. 124. Emphasis and text in square brackets added.
[164] Behe, M. 1996. *Darwin's Black Box.* Free Press.

random yet step by step way that Darwin imagined. We also know that almost all features of Living organisms are *specifically* (not randomly) complex and that *specific complexity* is only ever known to have intelligent causes.

'We are starting to realize that there's nothing in the laws of unguided, Life–Forceless, mineral Chemistry to cause the Life Forms to arise by 'accidental' and 'unintelligent' means any more than any of our vastly less sophisticated machines, computers and the complex codes to run them ever arise 'by accident'. It is, therefore, clear that subtle, superordinate Life Forces of great intelligence and complexity are responsible for the ongoing, second by second existence of all the Life Forms.

'We call them 'Life' Forms because they are imbued with a Life principle. Ideologically materialist science has claimed, since the artificial synthesis of urea in 1828, that there is no Life or Vital Force or Life Principle and, since Darwin, that Life is not a 'super' or 'higher' Natural phenomenon at all. This, however, does not work. Life is not a purely 'ordinary' or 'lower' Natural phenomenon and it cannot be explained away just in terms of randomly mixed chemicals having 'accidents' over a very long period. That is a truly extraordinary claim for which there has never been one thimbleful of extraordinary evidence.'

'This news, Alisha, while inspiring and exciting, just like the modern evidence for the scientific reality of various psychic phenomena and of Life after death,[165] is going to take our civilization a while to accommodate.

'Returning to Stephen Meyer and his scientific hypothesis that the genetic code is evidence for intelligent design; he says:

"The theory of intelligent design, like the other historical scientific theories it competes against, is tested against our knowledge of the evidence in need of explanation [the *intelligent digital information* in the DNA CODING] and the knowledge of the cause and effect structure of the world."

'He explains that scientists of Natural history, like Darwin, who cannot do lab experiments, *must choose between competing hypotheses* as to which has *"the best explanatory power"*. He says our

---

[165] Fontana, D. 2005. *Is There an Afterlife?: A Comprehensive Overview of the Evidence*. Schwartz, G. E. Phd. & W. L. Simon. 2002. *The Afterlife Experiments: Breakthrough Scientific Evidence of Life After Death*. Carter, C. 2010. *Science and the Near Death Experience*.

experience and knowledge of different kinds of causes (for example intelligent causes versus unintelligent and accidental ones) guide us as to the kinds of effects that they will generate.

'Frost, rain, snow, ice and rivers, etc are relatively unintelligent and random causes and they produce relatively unintelligent effects. They are the causes of *randomly* evolved river valleys and other landscapes. They are, however, 'causally inadequate' to create clever mechanisms or machines or Coding systems of any kind – not even wooden toys.

'Basically, Ali, in *Signature in the Cell*, Stephen Meyer is explaining that the DNA CODE, like all codes and other languages is a mental and symbolic device, *not* a material one, and that it cannot be attributed to *unintelligent* or purely material causes any more than any other code or software can – not Mozart's music notation, not Shakespeare's sonnets and plays, not ancient hieroglyphs, not Morse Code, none of them. It is only a biology taken hostage by the materialist world view that denies this.

'Meyer says: "Considerations of *causal adequacy* provide an experience-based criterion by which to test, accept, reject, or prefer competing historical theories,"[166] in this case the materialist ('accidents-create-codes'), unintelligent design theories *versus* the Idealist ('intelligence-creates-codes') ID theory. Meyer says that when a theory, like intelligent design, cites causes, such as intelligence, which are **already known to produce the effect in question,** such as digital codes, they meet the test of **causal adequacy** and when, like the materialist theories, they do not cite such causes, they fail to meet the test of causal adequacy. The laws of unguided chemistry are causally inadequate to create codes – it's not improbable, it's impossible. Meyer says:

> '… Like other historical scientific theories, intelligent design makes claims about the cause of past events, against our [real-world] knowledge of cause and effect. Moreover, because experience shows that an intelligent agent is not only a known, but also the ***only*** known cause of specified, digitally encoded information [like COBOL, BASIC, Morse code, and the digitized DNA CODING itself], the theory of intelligent design developed in this book has passed two critical texts: the tests of [1] *causal adequacy* and [2] *causal existence* …[167]

---

[166] Meyer, S. 2009. *Signature in the Cell*. Harper One; p. 405. Emphasis added.
[167] Ibid., p. 405. Emphasis and text in square brackets added.

'Meyer explains that precisely because ID theory, unlike the various materialist and unintelligent design theories for the origins of the DNA CODE uniquely passes the tests of (1) causal adequacy and (2) causal existence that he concludes that intelligent design:

'stands as the *best explanation* of the DNA enigma.'[168]

'Meyer is right. The only known causes of codes are intelligent agencies. But because materialist biology assumes existence to be purposeless and accidental, all its 'explanations' of the Living World then flow from that key but unproven dogma or prejudice. This is why it is so opposed to intelligent design thinking in biology because ID *refutes* the materialist, unintelligent design paradigm – it helps to restore meaning, Soul, intelligence and purpose to the Cosmos and that happy, exciting, and certainly very interesting conclusion, is something, strangely, that materialist thinking in science and biology seems to find inconvenient or depressing rather than interesting and exciting.

'Materialist thought, Alisha, sees existence as an unintelligent and mono–dimensional Mystery. Idealism thinks of it as an intelligent and multi–dimensional Mystery. In both these thought schemes, however, the impossible has happened. The only difference is that in one scheme it happens earlier, in the other later.'

'How do you mean, 'the impossible has happened'?'

'I mean that in both these schemes the miracles and magic of Life, Consciousness, Subjectivity, Sentiency, Awareness and Intelligence exist or have occurred. Yet why should they? It seems impossible. As impossible as the fact that even one quark or photon should exist. Yet, these 'impossible' phenomena of both Consciousness and Matter or Subject and Object, Witness and Witnessed, Mind and Form existing at all, amazingly, unbelievably are.

'In the materialist scheme it all happens unintelligently – first trillions of imaginary multiverses, then 'accidental', 'unintelligent' matter in this mindless and meaningless section of the multiverse we are in, then 'accidental' consciousness, mind and intelligence, no one has a clue how or why – which, in the materialist scheme, doesn't matter because there is no rhyme or reason to any of it, it's all just an 'accident'.

'In the Idealist scheme it's the other way round. Consciousness, Mind, Intelligence, Love and Purpose *simply are,* as ontological extants, no one has a clue how or why, for why should they *simply be,*

---

[168] Ibid., p. 405. Emphasis added.

yet so it is, and from this the universe we are actually in and all its mechanisms and Soulful Life Forms flow.'

'So all we can do, Ollie, is choose between these two equally 'impossible' mysteries in coming to our understandings of Life and existence?'

'Yes. Our only choice is between (a) a mindless and accidental 'impossible' Mystery of existence, or (b) an intelligent and meaningful 'impossible' Mystery of existence, for why should anything at all exist? – mindless or intelligent – it's 'impossible' isn't it? How can there be anything at all, rather than nothing at all? It is no less absurd that this surreal impossibility that anything at all should exist, rather than nothing at all, includes within it the astonishing 'impossibility' (or miracle) that it is intelligently and divinely or 'super' Naturally multi-dimensional, Conscious, Alive, Aware *and Meaningful* rather than unconscious, mindless and accidental!

'Our own amazing *intelligences, meaning discerning and meaning creating* capacities should be clues enough. Not clues enough, though, for poor Faust and his intellectual descendants. Mephistopheles, Faust's mentor and tormentor wants us to think existence is meaningless, Soulless and unintelligent, who knows why, but that is his wish.

'Materialist thought, as it studies the Living World, DNA CODING "more advanced than any software", tiny Living Cells more complex than Jumbo Jets, a flower blossoming, a butterfly hatching, the smile on a child's face, being a gloomy, glass half-full ideology opts for an 'accidental and meaningless' impossible–mystery of existence. Idealist thinkers, on the other hand, being more glass half-full types, study the same phenomena, and opt for the 'intelligent and meaningful' impossible–mystery of existence.

'Materialist thinking science and biology embrace the 'meaning' of existence by saying that there's nothing to embrace. It's all meaningless! We are all 'unconscious zombies' and 'nothing but accidental packs of neurons'. The Idealist scientific perspective, on the other hand, embraces existence by saying it's intelligent, it's something to love and be interested in, challenging as it is, and because it is *intelligent* it is, by definition, purposeful and *intelligible* which implies that wherever we look, be it at these grasses, the flowers, the animals in these fields, a tadpole, a caterpillar, a butterfly hatching, the people, our families, children playing, our friends, our pets, the wind waving trees, an eagle on the breeze, the sky, the sun, the moon and the stars beyond we see *intelligence* and *purpose* and Life and Soul and *meaning* reflected back. Idealism holds that it is because of these very qualities of the Cosmos that we are able to do science at all. This is why I'm passionate about this, Alisha. We live in a

Cosmos of meaning, intelligence and purpose not one of mindlessness, unintelligence and accident.'

'Stephen Meyers' great achievement in *Signature in the Cell*, though, is that using empirical, evidence–based arguments, he has *refuted* the materialist, unintelligent–design line and its 'mindless-and-accidental-mystery' theory of Life and existence. He has disproved the non-evidenced-based materialist claim that the *best* and *most causally adequate* explanation for the Living World is 'accidental chemistry'.

'A reviewer said of *Signature in the Cell*:

> 'Meyer demolishes the materialist *superstition* at the core of evolutionary biology by exposing its Achilles' heel: it's utter blindness to the origins of [the intelligent] information [that runs all the Life Forms]. With the recognition that cells function as fast as supercomputers and as fruitfully as so many factories, the case for a *mindless cosmos* collapses. His *refutation* of Richard Dawkins will have all the dogs barking and the angels singing.[169]

'Meyer's magisterial, 600 plus pages-long work shows that it is no longer intellectually tenable to argue that the digital DNA CODING, that runs all the Life Forms, is the dynamic and clever result of unintelligent and 'accidental' processes. It's not merely improbable, it's *impossible.*'

'And the significance of this, Ollie, is ...'

'It is incredibly significant because the *entire* Living World is run, at least in part, by the DNA CODE. Even those astonishingly complex, early bacteria, dating back 3.8 billion years, were run by this same clever, digital coding. This knowledge tells us, as if it wasn't obvious, that Living Nature is *intelligent* in origin and in its continuous, *second to second,* 24/7–clever, bio-functioning, as collective humanity has always intuited, as almost all the worlds great philosophers and scientists, from Plato and Aristotle on, always *rationally inferred* and as all the world's great spiritual teachers, the saints and the sages the prophets and the mages have also always maintained.

'Meyer and Dembski's work and the work of the other scientists and thinkers that they draw on tells us that there is no scientifically plausible way that the DNA CODE can possibly be the incredibly clever product of random 'chance' – the laws of unguided, accidental, entropic chemistry forbid it. The exciting news is that this amazing

---

[169] George Gilder, author of *Wealth and Poverty* and *Telecosm*; *Signature in the Cell.* Emphasis and text in brackets added.

CODE that helps to run all Living things, whether we consider it spooky and weird or magical and amazing, is not the random effect of unintelligent causes. It really is as clever and purposeful as it looks and it is clearly intelligently designed which, given that it is, as Bill Gates once said, "like a computer program but far, far more advanced than any software we have ever created", is no surprise.'

11 Material–Science: Brilliantly Explains How Things Work

Physical–Science is brilliant – it explains what things are 'made of' and how they 'work', what Aristotle called their material and mechanical causes. Spiritual–Science, by contrast, seeks to explain the ideas, purposes and meanings of things, what Aristotle called their formal (shaping) and final causes.

'One thing, Ali, I have huge respect for the achievements of modern physical–science. They are awesome. It has made possible the phones, computers, radios, TVs and Internet that we use all the time. It unraveled the mysteries of the DNA CODE and genomes. This kind of physical–science is a more or less philosophically neutral activity, it tells us how things 'work' and what they are 'made of'. It takes no position on the meanings and purposes of things, what Aristotle called their formal and final causes, the ideas and purposes behind them.

'Materialism, on the other hand, although it dominates science today, is not science as such but a belief system. Materialism believes that 'only matter is real' and that Life is, fundamentally, mono-dimensional and meaningless because, according to this depressing ideology, Darwinian random evolution has proved the Living World to be an 'accident'. Ideological–materialism asserts that science and materialism are one and the same thing or, at the least, that when doing science we must assume that all things have purely material explanations. This is methodological–materialism. It is also sometimes known as methodological–naturalism, which is the idea is that all things have purely Natural explanations.

'That last part sounds reasonable enough.'

'Yes, it does but it's a deceptive phrase because 'Nature' is, obviously, whatever exists. If Nature is 'mindless, unintelligent and accidental' then whatever exists must be 'mindless, unintelligent and accidental'. If Nature is 'intelligent' and 'purposeful' then whatever exists must be intelligent and purposeful. This, of course, is at the heart of the dispute between ideological–materialism and philosophical–idealism. What is the true nature of Nature? Is it material only and composed of synergistically nested hierarchies of apparently (only) intelligently functioning 24/7 'mindlessness', 'unintelligence' and 'accidents' all the way up and all the way down or does it combine in itself Consciousness, Mind, Intelligence and Materiality?

'In materialism Nature is considered to be mindless, unintelligent and mono–dimensional. This means that humanity's witnessing awareness, sentiency, thinking, feeling, love, purpose and intelligence are all regarded as 'accidents' of atoms of carbon, oxygen, hydrogen and nitrogen plus a few other chemical elements. Idealism, on the other hand, sees total Holistic Reality as intelligent and multi–dimensional. It takes seriously the classical rational inference and widespread human intuition that there are dimensions and vibrations of existence, not accessible to physical instrumentation, whose existence can be rationally inferred, which can, also, be explored in consciousness by inner means[170] and to which we meta–morphose and trans–form on 'death'. Materialism, on the other hand, denies that consciousness survives the death of the chemical body and refuses to properly study 150 years of accumulated evidence and research that says otherwise.[171]

'A science dominated by materialistic thinking *does not to wish to know* about the large body of modern evidence for the truth of psychic and spiritual phenomena of all kinds, including the evidence from OBE's and NDE's (out of body[172] and near death experiences[173]) and that consciousness can exist apart from the body.[174] Far from being curious and open minded, this kind of materialist scientific thinking is hostile to the evidence for Life after death. Because they don't match its paradigm it is also hostile to the, by now, voluminous evidences for extrasensory perception, for vibrational and energetic healing which involve the subtle, non-physical frequencies and dimensions of holistic Nature which materialism refuses to recognize.

---

[170] Peake, A. *The Out of Body Experience: The History and Science of Astral Travel.* Watkins Publishing. Monroe, R. A. 1972. *Journeys Out of the Body.* Souvenir Press. Yogananda, P. 1946. *Autobiography of a Yogi.* Self-Realization Fellowship. Gustus, S. 2011. *Less Incomplete: A Guide to Experiencing the Human Condition Beyond the Physical Body.* O Books.

[171] Fontana, D. 2005. *Is There an Afterlife?: A Comprehensive Overview of the Evidence.* O Books. Review, product description, Amazon.

[172] Monroe, R.A. 1972. *Journeys Out of the Body.* Souvenir Press. When, out of the blue, Monroe began to have out-of-body experiences, he was alarmed and incredulous. He found himself leaving his physical body [and able to explore] a world unbounded by time or death. As he met others [familiar with] the experience and read the long history of those who have known the phenomenon in the literature of the East, his fears were alleviated, ... the effects changed his life. ... Monroe challenges us to rethink our ideas about life and death and, with step-by-step instructions, invites us to initiate the OBE experience for ourselves.

[173] Lommel, P. 2010. *Consciousness Beyond Life, The Science of the Near Death Experience.* Harper One. Fenwick, P. & E. Fenwick. 1995. *The Truth in the Light: An Investigation of Over 300 Near Death Experiences.* Headline Book Publishing.

[174] Kelly, E. F. & E. W. Kelly, A. Crabtree, A. Gauld 2009 *Irreducible Mind: Toward a Psychology for the 21st Century.* Rowman and Littlefield.

'Chris Carter's book, *Science and Psychic Phenomena: The Fall of the House of Skeptics* illustrates that these attitudes in mainstream, materialist science are held very dogmatically and almost amount to a kind of pseudo-religion. He cites carefully controlled modern research showing that telepathy, remote viewing, precognition and psychokinesis are real and that skepticism of such phenomena is a materialist prejudice – it is not based on hard science.

'He explores the documented, *reproducible evidence* that such abilities are real. The mainstream scientific community, he explains, has, however, vehemently denied the existence of psychic and spiritual phenomena for a very long time now. The reason for this is not that extrasensory perception and other 'super' normal abilities cannot be verified but because their existence challenges cherished dogmas based in materialist and sometimes even religious beliefs – not hard science. The skepticism persists partly because mainstream science, captured by materialist thinking, *has become closed minded* and simply refuses to be interested in or to look at the, by now, vast body of modern evidence proving beyond any reasonable doubt that psychical and spiritual phenomena are real and that consciousness is not an 'accident' of matter. Such a science is no longer open-minded but has become a vehicle for a new kind of materialist faith system whose unreasonable and non-evidence-based gods are Mindlessness, Unintelligence and Accidental Chemistry.

'This is a science that insists that 'Life-is-accidental-chemistry' with no real evidence for this extraordinary claim other than the materialist assertion that *it must be so* because the Cosmos, materialism believes, is mindless and accidental and so, therefore, must be Mind and Life!

But that seems self-refuting, Ollie. Either matter itself is, somehow, imbued with mind and intelligence, or, mind and intelligence are principles apart from matter, yet able to meaningfully and purposefully combine with it.'

'I agree, Ali, and, in any case, science ceases to be science if it becomes an ideology. Scientific beliefs, such as the materialist concept, for example, that 'only mindless matter is real and it accidentally gives rise to intelligent minds' should not be held as dogmas but, to be genuinely scientific, must be hypotheses and theories that are:

'open to testing and hence possible rejection.'[175]

---

[175] Carter, C. 2010. *Science and the Near Death Experience.* Inner Traditions; p. 237. Emphasis and text in square brackets added.

'How can science be an objective methodology for discovering truths about reality if it is dogmatically committed to various metaphysical beliefs that cannot be questioned and which, therefore, exclude certain kinds of evidence simply because the evidence contradicts the metaphysical belief?

'This reviewer of *Science and Psychic Phenomena: The Fall of the House of Skeptics* said:

'Like the title suggests, this book is two stories intertwined, one charting the scientific discovery of psychic powers (psi) over the last century and another castigating a misguided social movement known as [materialist] skepticism for claiming to know better. Chris Carter surveys the sea of anecdotal and *statistical evidence* for the existence of telepathy, clairvoyance (also known as remote viewing), precognition and psychokinesis. Skeptics, meanwhile, maintain that psi [extrasensory ability and 'paranormal' phenomena] is incompatible with what we know about reality and therefore must be false. Yet *psi phenomena do not violate any known principles of physics.*

'Rather than face the evidence head on, self-proclaimed skeptics are engaged in a holy war, says Carter, "fueled by the fervent belief that they alone are the last defenders of the citadel of science." As to real scientists, most do not identify with organized skepticism. Going all the way back to Herodotus, Carter examines the history of psi, including the findings of the Society for Psychical Research, J. B Rhine, Daniel Home and Charles Honorton, whose 'autoganzfeld' procedure was immune to charges of human tampering. He also discusses statistician Jessica Utts' claim that "psychic functioning has been well established" by ordinary scientific methods.

'Carter contrasts the sober science of psi with the crusading fanaticism of the Committee for the Scientific Investigation of Claims of the Paranormal. CSICOP, an organization straight out of Orwell, completely avoids scientific investigation. James Randi, Richard Wiseman, Susan Blackmore and Ray Hyman all get singled out for extensive scrutiny. Needless to say, their methodology is found wanting.

'In the face of skeptics who claim that all research into psi [psychic and spiritual phenomena] is pseudoscience, Carter charges that ideological skepticism represents a mutant form of science known as scientism, which is more concerned with absolute truth than such banalities as *hypothesis, experimentation and theory*. The only skeptic who emerges from Carter's analysis with a shred of integrity is [Susan] Blackmore, who at least concedes she was biased and might have got it wrong.'[176]

In *Randi's Prize: What Sceptics Say About the Paranormal, Why They Are Wrong, and Why It Matters*, Robert McLuhan, also debunks the arguments of the debunkers and materialist hyper–skeptics such as James Randi. McLuhan examines the disproportionate influence of high-profile materialist 'skeptics' in shaping public and scientific opinion about such things as telepathy, extra-sensory ability, near-death experiences and Life after death. A scientist reviewer says:

'If you haven't investigated the scientific research regarding psychic phenomena objectively ... *Randi's Prize* is an excellent place to start. ...

'Yes, psi effects are elusive but, as *Randi's Prize* makes clear, they *have* been studied in painstaking detail for 150 years by highly qualified scientists, including, in recent decades, **some of the most carefully executed scientific experiments ever conducted** with multi-layered experimental controls that put other fields of science to shame. Because researchers in this field are under unrelenting, often vicious assault, they control even against absurdly improbable and unrealistic forms of cheating and fraud among other things, problems that most scientists don't have to think about at all (imagine trying to work in that environment). Statistically and **taken as a vast body of work, their results are rock solid** ...

'If you are new to the topic of psi as understood through science this is a fascinating overview of a noisy controversy that has largely been manufactured by a relatively small number of extreme 'skeptics'. Both sides of this polarizing issue are treated sympathetically and fairly – in my view McLuhan shows amazing restraint and civility when dealing with examples of seemingly blatant intellectual dishonesty [on the side of the skeptics] ...

'The scientific facts are in; they're well-proven and extensively documented – many tens of thousands of pages of detailed studies. ***The demand for more proof is simply a ploy.*** Robert McLuhan's book makes it clear that what has largely been missing for the past century is *a fair and civil debate on the actual, undistorted, scientifically demonstrable facts.* This book sets a good example ...'[177]

---

[176] Ted Dace, back cover, *Parapsychology and the Skeptics,* 2007, republished, 2012, as *Science and Psychic Phenomena: The Fall of the House of Skeptics.*

[177] McLuhan, R. 2010. *Randi's Prize: What Sceptics Say About the Paranormal, Why They Are Wrong, and Why It Matters.* Matador. Amazon reviews by Geophysics Ph.D, Sun Dog, M. R. Barrington and Robert Perry. Emphasis and text in square brackets added.

Material–Science: Brilliantly Explains How Things Work 155

'M. R. Barrington said:

'Robert McLuhan starts by listening to the psi-deniers, who sound plausible ... But [he] shows skillfully and convincingly that the [materialist] opposition is superficial, ill-informed, misleading, *sometimes blatantly dishonest* and occasionally bordering on the hysterical, as if afraid of the implications – which are indeed formidable. If you read just one book this year make it this one.'

'Robert Perry said:

'I thoroughly enjoyed Randi's Prize. I couldn't put it down, in fact. McLuhan's method is to walk you through his own mental journey, in which he was initially quite influenced by the [materialist] skeptics of parapsychology ... [until] he discovered ... that they systematically failed to really engage with their subject matter, keeping their distance from doing actual research and even from becoming genuinely acquainted with the original literature. ... with the sole intention of destroying it ...

'In the process, McLuhan essentially turns the tables on the debunkers, deconstructing their methods and practices under the spotlight of detailed examination, just as they purport to do with parapsychological research – a refreshing reversal indeed.'

'It is a non-negotiable spirit of materialism that makes mainstream science so dismissive of 150 years of accumulated psychic research and of the world's religious and spiritual traditions because of its own dogmatic faith system, that it is not prepared to question. Materialist science assumes that these traditions are relating to subtle, multi–dimensional realities that do not exist, that are purely imaginary. In fact, ideological–materialism sees science almost as a positively anti-supernatural, a positively anti-religious activity.'

'Isn't that putting it a bit strongly, Ollie? Surely science is simply about what's true? About testing hypotheses and scientific ideas and seeing where the evidence leads?'

'Yes, science should be like that. But what many people perhaps don't realize is that a science that has been captured by ideological-materialism is no longer an open minded science because it has already decided that 'only-matter-exists' – (even though matter, it turns out, is not really 'matter' at all but simply a form of intelli-gently informed, quantum energy s—p—a—c—e) – and it firmly opposes any and all evidence that suggests existence is intelligent and multi-dimensional. Its attitude is that the purpose of science is to explain *all* of Nature, including even the Living World, mono-dimensionally, by reference to purely material, that is entirely unintelligent and accidental processes ('just accidental chemicals + chemical accidents' to explain all) with no rhyme, reason or purpose to any of them. Remember Lewontin's definition of science?

> 'Our willingness to accept scientific claims that are against common sense is the key to an understanding of the *real struggle* between [materialist] science and the supernatural. We ... have *a prior commitment,* a commitment *to materialism.* ... we are *forced* by our **a** *priori* adherence to material causes [alone] to create an apparatus of investigation and a set of concepts that produce *material explanations* ... Moreover, that **materialism is absolute**, for we cannot allow a Divine foot, [or *intelligent* causation], in the door. The eminent Kant scholar Lewis Beck used to say that anyone who could believe in God could believe in anything.'[178]

'Well, that does show an absolute commitment to the materialist view of Nature.'

'Yes, it does. You see, for such thinking 'science' *is* materialism – a worldview that is implacably opposed to the idea that *anything* in Nature, in the Living World, is *intelligent* in origin and which insists that Life, Mind and Intelligence are simply accidents of mindless atoms and that no other explanations are possible.'

'But isn't that like insisting that TV Shows are, 'somehow', made of the metals and wires and plastics in the TV as opposed to merely transmitted by them?'

'Yes, it is. The trouble is, Ali, so many words, in this area, are freighted with so much psychological baggage. I don't know what Lewontin means when he refers to 'god' or the 'divine' but I wouldn't be surprised if he has some kind of a simplistic, medieval, bearded-god type agency in mind that no rational person believes in or needs to believe in. It's certainly not the kind of supreme intelligent Beingness I believe in. For me, 'God' and the 'Divine' are just one of several special words that people use to refer to the intelligent, multi-dimensional and *meaningful* nature of the total Holistic Reality in which we find

---

[178] Richard Lewontin (1997) Billions and billions of demons (review of *The Demon-Haunted World: Science as a Candle in the Dark* by Carl Sagan, 1997). *The New York Review*, January 9, p. 31. Emphasis and text in square brackets added.

ourselves – words like God, Brahman, Allah, Supreme Being, Buddha Nature, the Tao and so on.

'These words all imply a multi-dimensional Cosmos of intelligence and Meaning, to which we are intimately and integrally connected, not only materially and energetically but also psychically and spiritually, and that all things that exist are expressions of these multi-vibrational realms of universal consciousness, sentiency, intelligence and purpose. This is not intended to imply, as some religious and spiritual thinking may do, that we live in a Cosmos of perfection, not at all, simply one of *intelligence, purpose and ultimate meaning*.

'Materialist thinking dismisses this but if the modern spiritual-science researchers are correct and (a) there is Life after death and (b) reincarnation is a spiritual reality,[179] meaning that we experience all kinds of lives, and the opportunity to grow in wisdom and consciousness as a result, then the fact that temporary outer existence can be so harsh, difficult and, at times, so apparently meaningless, becomes tempered by this understanding.

'Materialism believes that people make up stories about the spiritual multi–dimensions, Life after death and reincarnation because they are scared of dying. However, an out of body or near death experience is not an idea or a belief it's an *experience*. One that often changes the experiencer's Life forever, frequently bringing great personal growth and transformation in its wake – greater kindness and compassion, sometimes previously unknown psychic and spiritual capacities and a complete loss of the fear of death.[180] There is now, extensive research into out of body and near death experiences.[181] So why there remains, among some, such a materialistic passion to rubbish all such evidence for the, ultimately, super Natural Cosmos of meaning and intelligence in which we exist as multi-dimensional beings of both Space and Time and beyond Space and Time is a mystery. It's time for our culture to be

---

[179] Stevenson, I. 1980. *Twenty Cases Suggestive of Reincarnation. Second Edition, Revised and Enlarged*. University of Virginia Press. Stemman, R. 2012. *The Big Book of Reincarnation: Examining the Evidence that We Have All Lived Before*. Hierophant Publishing.

[180] Grey, M. 1985. *Return From Death: An Exploration of the Near Death Experience*. Atwater, P. M. H.. 2010. *Coming Back to Life: Examining the After-Effects of the Near-Death Experience*. Transpersonal Publishing.

[181] Kelly, E. F. & E. W. Kelly, A. Crabtree, A. Gauld 2009. *Irreducible Mind: Toward a Psychology for the 21st Century*. Rowman and Littlefield; product description. Lommel, P. 2010. *Consciousness Beyond Life, The Science of the Near Death Experience*. Harper One. Fenwick, P. & E. Fenwick. 1995. *The Truth in the Light: An Investigation of Over 300 Near Death Experiences*. Headline Book Publishing.

skeptical of such exaggerated skepticism.[182] What are the extreme skeptics so afraid of? Why are they so afraid that there might just be 'more things in heaven and earth ... than are dreamt of' in their de-purposing philosophies of mindlessness, accident and unmeaning?

'Perhaps it is because for some thinkers concepts such as God or the Higher Natural all seem to imply absurd and arbitrary agencies who may or may not interfere with the Laws of Nature at any given moment. This view assumes that *only* the laws of Nature that it currently knows of exist and that there can't be any other laws of Nature, laws of Being, of 'what is', apart from the ones that it is presently willing to recognize or to infer. For example, the laws governing psychic phenomena, telepathy, remote viewing and spiritual mediumship demonstrating the survival of consciousness beyond physical death.[183] Just because we do not know how telepathy works doesn't mean it isn't a law based phenomenon.

'Materialism simply assumes in a dogmatic way that there *are* no 'higher' or 'super' Natural Laws that exist in parallel with and which at times seemingly conflict with or override the 'ordinary' Natural Laws that we take for granted or assume to be the only laws that exist. Clearly when people 'die', in 'ordinary' Natural terms, as opposed to Higher Natural or multi-dimensional terms, they are gone for ever, never to return in that particular physical image.

'Yet, few, perhaps, are aware of what might be called the 'trade secret' of modern biology which is that it has no idea what very 'Life' itself, that seemingly disappears on death, actually is. How can it? If it is in a "struggle with the supernatural" it can hardly admit that the most obviously 'super' Natural phenomenon of all is very Life itself.

'This is why we have to live with the bizarre fiction in our mainstream scientific culture that Life is not a principle in its own right, a force that is apart from 'dead', mineral matter, something more than chemicals that are 'accidentally' behaving in a Bio–logical or Life–logical way. And, absurdly, this view is based partly on the fact that someone artificially synthesized urea in 1828 and it was decided that this mean't that there was no Life Principle, that there were no Life or Vital Forces! This decision to conceptually abolish the Life Force and, subsequently, the Soul Principle is a major factor in the intellectual

---

[182] Carter, C. 2012. *Science and Psychic Phenomena: The Fall of the House of Skeptics.* Inner Traditions. McLuhan, R. 2010. *Randi's Prize: What Skeptics Say About the Paranormal, Why They Are Wrong, and Why It Matters.* Matador.

[183] Fontana, D. 2005. *Is There an Afterlife?: A Comprehensive Overview of the Evidence.* O Books. Review, product description, Amazon.

muddle that ideological-materialism has got us into as multi-dimensional Beings of true Life, Mind, Soul and Spirit.

'The late Dr Ian Stevenson, was a Canadian biochemist and professor of psychiatry. Until his retirement in 2002, he was head of the Division of Perceptual Studies at the University of Virginia School of Medicine, which investigates various super Natural phenomena. 'Stevenson theorized that the concept of reincarnation might supplement those of heredity and environment in helping modern medicine to understand aspects of human behavior and development. He traveled extensively over a period of 40 years to investigate 3,000 childhood cases that suggested to him the possibility of previous or past lives.'[184] These were cases of children who were talking of their recently ended previous life, describing people, places, events and the manner of their deaths which, in some cases, it was possible to cross check and verify. As one reviewer said of Stevenson's book, *Twenty Cases Suggestive of Reincarnation:*

> '"Suggestive" is putting it mildly (but is typical of Dr. Stevenson's cautious approach). "Pretty darn convincing" is more like it. This is one of Stevenson's original works from many years ago (the 1960s), but it has stood the test of time. If you read it and his more recent *Where Reincarnation and Biology Intersect,*[185] you'll have a fairly good idea of how convincing the evidence for reincarnation can be. Stevenson's research is very thorough, and this is a dense volume that can be a little dry to read. Each of the 20 cases is presented in sufficient detail to be compelling. It's not something you'll polish off in a couple of evenings, but you'll know you're in the presence of an honest-to-god researcher. Stevenson, who has been affiliated with the University of Virginia for decades, is almost single-handedly responsible for reincarnation being taken seriously in this country, and this book is a classic that should be one of the very first you buy on this topic. You'll be handing it out to skeptical friends and saying "Oh, yeah? Read THIS."[186]

'Roy Stemman is another reincarnation researcher. On a trip to Lebanon, he describes a photograph he took of a bearded, elderly man, wearing a white fez and a young girl seated next to him her hand

---

[184] Author's page, Amazon.
[185] Stevenson, I. 1997. *Where Reincarnation and Biology Intersect.* Praeger.
[186] Amazon USA

resting affectionately on his arm who looks as though she might be his granddaughter. It is not, however, an image of a granddaughter visiting her grandfather. Stemman's picture, both families believe, shows Ajaj Eid and his former, deceased wife Safa Eid (now reborn as Haneen Al-Arum a bus driver's eleven year old daughter). Safa Eid died on January 17, 1984 when the US battleship New Jersey began shelling the Druze positions in the Chouf mountains above Beirut. Safa followed her husband outside to warn him to be careful as he went to help a neighbor whose curtains had caught fire in the detonations. Stemman says:

'Still able to recall that day, Haneen – who was 11 years old when I met her at the spot where Safa's life ended – told me what happened next. "There was more shelling and a piece of shell hit my neck and I died."

'As soon as she could speak after her rebirth as Haneen, three years later, she began calling for her son, Riad. She told her parents, "I am Safa," adding that she was from Bchamoun and the Eid family. In a culture that accepts reincarnation, many go in search of family members that might have been reborn, and Shahira, one of Safa Eid's daughters, was no exception. She learned of three young girls who might be her mother's reincarnation, one of whom was Haneen. When they met and she asked the young girl, "Do you recognize me?" Haneen replied, "You are Shahira, my daughter," and they hugged and kissed. I saw for myself how very close Haneen was to all the family members when I accompanied her to the Eid home. When first taken there by car I learned, Haneen had pointed immediately to the house where the curtains caught fire, as she stepped from the car. Shahira, still testing Haneen to be sure she was her mother's reincarnation, started to walk up some stairs as if heading for the family home, but the young girl corrected her, pointing down the hill, instead. Shahira changed direction and entered the first gate. "This is not our house," Haneen protested. When they reached the right house, she remarked that the front had changed [which was correct as it had been damaged in the shelling that took her life as Safa]. ... [Once inside her old home she pointed out that her former husband now had a beard where before he had been clean shaven and] she stopped at a display cabinet and remarked that she had hidden money in the lower compartment for Layla [her former daughter in law]. She was clearly disappointed to see it had now gone. In fact, Layla confirmed to Riad that she had found it and removed it. Layla also had a test question for Haneen: "Is that where you kept your gold?" she asked, referring to the compartment. "I never had gold," Haneen [correctly] responded.'[187]

'That's an interesting account, Ollie.'
'I agree, it is and there are many such accounts, documented thoroughly in the late Dr. Ian Stevenson's life's work. You see, if Life and Soul are principles apart from matter, in a multi-dimensional and Holistic Nature, then 'death', it may turn out, as the bulk of humanity has always realized, is but a dimensional transition of consciousness – the specific TV (the chemical body) may have worn out but the TV Show (the Mind and Soul) continues elsewhere and, it seems, can also return to this dimension. Closed-minded materialist thinking dismisses such evidence because it regards it as 'super' Natural and it dogmatically maintains that there is no such thing. Equally, some closed minded religious views will dismiss such evidence and seek to explain it away.

'Yet the distinction between the so-called 'ordinary' Nature, consisting in chemical matter and the many mechanisms it builds (which materialism believes in) and so-called 'super' Nature (which materialism denies) – pre-existing Mind or Minds, the self-existing principles of consciousness, sentiency and intelligence – is an artificial and manmade one. It is we who label parts of Nature 'ordinary', take them for granted, nothing-special-Natural and other parts of Nature 'super' and miraculous or – if we're materialists – completely nonexistent and imaginary.

'Yet, is, for example, supposedly 'ordinary', take-it-so-very-forgranted, it's-nothing-special-at-all gravity really just so very 'ordinary' rather than 'super'? It only stops the entire Universe from floating apart! That seems pretty super-functional and teleological to me, Ali. Materialist science, however, believes that gravity is an 'accident'. How does it know? It doesn't know. It doesn't have a clue why gravity exists. Just because we know how it 'works', to some extent, doesn't tell us *why* it is. Saying it arose after the Big Bang is only to **describe**, it is not to **explain**. It is, at best, a partial or half-explanation, encapsulating Aristotle's first two causes, those of matter and mechanism but ignoring his last two causes, formal and final, the ideas and purposes of things.

'Describing a Big Bang is not the same as saying *why* there should be one in the first place. A question to which, in essence, there are only two possible answers – (a) a mindless accident of a Cosmic Mystery, as

---

[187] Stemman, R. 2012. *The Big Book of Reincarnation*, Hierophant Publishing.

materialism believes, or (b) an *intelligent purpose* of a Cosmic Mystery as philosophical idealism, collective humanity and the worlds religions and spiritual traditions hold.

'It is only our *attitude* and psychological perspective that causes us to say: 'Oh gravity, it's not *super* Natural, it's just ever-so *ordinary* Natural! It's just there, there's nothing special about it.' Alternatively, we might say: 'Gravity is a miracle! It makes the entire material universe possible'. The same goes for our attitudes to: (1) 'it's nothing special, it's an accident magnetism' or (2) 'it's amazing and magical magnetism,' or to (1) 'so what, it only powers everything, including our bodies, but it's just an accident electricity,' or (2) 'it's amazing and exciting electricity,' (1) 'boring it's nothing special, it's an accident oxygen,' or (2) 'it's awesome oxygen', (1) 'it's nothing special, it's an 'accident' carbon,' or (2) 'it's amazingly versatile carbon,' the same for sultry sulphur, fiery phosphorus, numinous nitrogen, heavenly hydrogen and all the other smart-materials-chemicals which we can regard as (1) 'meaningless accidents' or (2) 'intelligent miracles' – for none of us knows why any of these should be. Saying 'a Big Bang' or the 'multiverse' did it is about as pathetic as it gets by way of explanation. *It is not a full explanation, it's but a description.* It's nothing more than saying 'this happened, then that happened' and but a description of matter and mechanism.

'Materialism attempts to deal with this by resort to its unreasonable (and impossible) knowledge–gap filling pseudo–uber–gods of Mindlessness, Unintelligence and accidental chemistry. In true Mind, Life and Soul denying materialism there can be no meaning. Total Natural Reality is just meaninglessly and accidentally there. And as soon as there's a hint that total Reality might be stunningly intelligent, based, for example, on the stupefyingly precise laws of physics, materialism's dark church invents the idea that universes are arising more or less every second in such a way as to make the universe we are actually in totally meaningless. Phew, what a relief, the Cosmos is utterly pointless after all! This kind of unreasonable pseudo-science answers none of the questions that really matter, those relating to meanings and purposes.

'It doesn't explain *why* there should be a Big Bang resulting in amazing electricity, ineffable light, astonishingly smart-materials-chemicals, bio-molecules of astounding complexity, DNA CODING "more advanced than any software" we've ever created.

'In the real-world those questions, in essence, have only one of two possible answers: (1) Materialism's answer is a mindless and unintelligent, impossible Mystery of Existence, ('impossible' because why should anything exist at all), woven out of mono-dimensional unconsciousness, impotence, unintelligence, accident and unmeaning or (2) Idealism's answer which is an intelligent, multi-dimensional and purposeful Mystery of Existence, woven out of Consciousness, Sentiency, Purpose, Love, Intelligence and all the astonishingly complex mechanisms of outer Nature.

'How we view Holistic Reality is, in the end, not a matter of 'science' but one of attitude. We all believe what we want to believe. No one can even 'prove' to us that the moon isn't made of cheese if we don't wish to believe their proof. Some use 'science' to support their view that existence is meaningless and that our amazing Living World is an 'accident'.

'Others, more glass half-full types, call science in favor of their view that Nature is stunningly intelligent, extremely 'super' at all levels dimensions and vibrations and not 'ordinary' at all. The DNA CODING is not 'ordinary', it's truly amazing, but in the materialist glass half-full view of existence it 'must be' an accident because 'isn't everything?' You see, Ali, what is in dispute, all the time, in the conceptual debate between ('accidents-create-all') materialism and ('consciousness-and-intelligence-before-all') idealism, is the very nature of Nature itself.'

'But surely, Nature is just what ever exists?'

'Yes, of course, but the very dispute is as to *what does exist?* The materialist school firmly believes that Nature is all accidental, purposeless, [which is self-refuting] mindless, solid, stolid, tiny billiard ball atomic matter and that there is no inherent, ontological *consciousness, mind or intelligence anywhere to be seen* – just accidental us and some accidental animals. Philosophical Idealism says, on the other hand, that consciousness, mind and intelligence are paramount. Philosophical idealist, Max Planck, discoverer of quantum physics said, for example:

> 'As a man who has devoted his whole life to the most clear headed science, to the study of matter, I can tell you as a result of my research about atoms this much: *There is **no matter as such***. All matter originates and exists only by virtue of ***a force*** which brings the particle of an atom to vibration and holds this most minute solar system of the atom together. We must assume behind this force the existence of a ***conscious and intelligent mind***. This mind is the matrix of all matter.'[188]

---

[188] *Das Wesen der Materie* [The Nature of Matter], speech at Florence, Italy (1944) (from Archiv zur Geschichte der Max-Planck-Gesellschaft, Abt. Va, Rep. 11 Planck, Nr. 1797). Emphasis added.

'So if matter, ultimately, arises out of and is sourced in a self-existing, self-arising matrix of Consciousness, Mind and Intelligence, as the yogis and the world's indigenous peoples, shamans, mystics and spiritual teachers have also always held, then the whole thing, the entire cosmos, from sub-atomic quark to galaxy, is, ultimately, super Natural and is, in certain senses, literally alive, sentient, intelligent and meaningful and composed of many dimensions, vibrations and frequencies governed by many diverse laws, forces and processes. Some of these laws and forces we understand well and others less so.

'So, for example, if some usually reliable and trustworthy witnesses attest to the seemingly *higher* Natural phenomenon of levitation, as has occurred from time to time, we have a choice: Our current physical–science understanding is that gravity is a universal law. How could it ever be gainsayed? The skeptical materialist view is: 'No, levitation is impossible. Gravity is an inviolable *ordinary* Natural Law. So-called levitation is always a *trick* of some kind. Those usually sober and reliable witnesses who attested to the levitation episode in question must have been drinking or smoking something that clouded their usually reliable judgment.'

'What such materialist thinking is really saying is that it 'knows', for a certainty, that *only certain* Natural Laws exist and that is that. Surely, it would be so much more scientific and open-minded to say that:

> "Clearly, levitation is not something that happens every day. In fact, if it happens at all, it is obviously extremely rare. Yet, there *are* accounts of it, throughout history, and they persist. Although this phenomenon doesn't fit in with our *current* understanding of the laws of physical reality, perhaps it is, nonetheless, possible. It may be that the sub-atomic or quantum-energy-field of the person levitating so changes that they temporarily become lighter than air, almost weightless, and so they are able, briefly, to rise above the ground. This provides a possible theory for levitation. No laws have been violated – just unusual ones activated."

'Spiritual adepts like advanced yogis and saints who have performed or experienced these kinds of unusual events, such as levitation, have always claimed that everything they do or experience is not arbitrary, it is *as much subject to laws as anything else*. We walk and are subject to the 'natural' laws of gravity. A yogi or saint[189] levitates or a spiritual

---

[189] St Theresa of Avila is said to have levitated during mass. St. Joseph of Cupertino reportedly levitated for extended periods on many occasions. Yogananda cites examples of levitation in his famous book, *Autobiography of a Yogi*. The Buddha is

master walks on water under the aegis of other 'natural' laws whose workings are currently not understood. If the entire cosmos is, as the yogi–adepts of India said centuries ago and as classical, western, scientific–Idealist thinking has always maintained, ultimately an expression of consciousness, force, frequency and vibration, that is, fundamentally, a mental place and someone is a genuine adept, a spiritual master, then it is not so absurd that, as a matter of will, they could so alter their quantum vibrational rate that they literally become as 'light as a fleck of goose down', and therefore able to levitate. Because this is extremely rare, however, materialist thought believes it simply can't happen, rather than keeping an open mind.

'People, including scientists, think seriously about the idea of *teleportation,* which seems pretty far out to me, Ollie, so, in comparison, levitation doesn't seem such a stretch.'

'I agree. Listen to this:

> 'Daniel Douglas Home was said to repeatedly defy gravity over a career of forty years. … He could also cause tables and chairs to rise feet into the air, and was never demonstrated to be a fraud by hundreds of purportedly skeptical witnesses, except one. … Home's fame grew, fueled by his feats of levitation. Physicist William Crookes claimed to have observed more than 50 occasions in which Home levitated, many of these at least five to seven feet above the floor, "in good light."[190] More common were feats recorded by Frank Podmore: "We all saw him rise from the ground slowly to a height of about six inches, remain there for about ten seconds, and then slowly descend."[191] One of Home's levitations occurred in 1868. In front of three witnesses (Adare, Captain Wynne, and Lord Lindsay) Home was said to have levitated out of the third story window of one room, and in at the window of the adjoining room.'[192]

'But what does this prove, Oliver? So what if some people can levitate? What's so miraculous about it? Is it more miraculous than Life itself? Is it more miraculous than the bulbs in the ground that emerge

---

said to have walked on water, and, of course, also Christ. Yogi, Subbayah Pullavar was reported to have levitated for four minutes in front of a crowd of 150 witnesses on the 6th June, 1936 (Wikipedia).

[190] Sir Arthur Conan Doyle. 1926. "The History of Spiritualism" volume 1, p. 196.
[191] Podmore. 1903. *Mediums of the Nineteenth Century, Part 1.* Kessinger Publishing, (2003); p. 254.
[192] Doyle, Sir A. 1926. *The History of Spiritualism, volume 1;* pp. 196–197 (Wikipedia).

into beautiful bluebells and daffodils in the spring, no one truly knows why, especially as materialist thinking science denies the Life and Soul Forces? Is it more miraculous than a child being born or a butterfly emerging from the chrysalis' tomb? Surely we are a dead civilization if we need demonstrations like levitation to convince us that Life and Soul and Existence are the greatest miracles of all. The appearance of an entire universe from nowhere to now here!

'Ali, you're right. If we have eyes to see and ears to hear we notice that we are surrounded by miracles all the time. Light and electricity are miracles. Atoms and molecules are. Photosynthesis is a miracle and the carbon cycle. Water and its amazing and unique life-supporting properties different to all other liquids. Breath is a miracle, and metabolism. Sexual reproduction. Mind and intelligence. The *minds and intelligences* that can conceive and manifest Rolls-Royces or Supercomputers (both so much simpler than the Life Forms) are miracles. The transformation of a caterpillar into a butterfly or a tadpole into a frog is a miracle – one in no way diminished by our capacity, today, to cleverly describe some of the bio-chemical processes involved, their material and mechanical causes. Even so, Ali, so-called miracles involving people are interesting in that they have the potential to wake us up to the extraordinary in a more challenging and immediate way.

'For example, most people need to eat to live. Whether they can last for four weeks without food, a feat readily possible for many, or for a couple of months, there comes a point, for most of us, that without physical food we perish. As with the intriguing phenomenon of levitation, there are, however, accounts, down the ages, of people, right up to and including today, who either eat nothing at all, sometimes even for decades, or who eat very little. Such accounts, to materialist thinking, once again, violate inviolable Laws of Nature – so they are discounted, mocked or ignored. An ideologically materialist science that is "in a struggle with the supernatural" (although what is more super Natural than very existence and the fact that there should be even just one photon rather than nothing at all) will dismiss all such accounts as so much make believe and hokum. Be that as it may, listen to this:

> 'In 1923, Therese Neumann, a nun in Southern Germany, stopped eating and drinking. Apart from the wafer given at Mass, she did not eat again, despite living for a further 35 years. Other similar cases have been reported over the years – often holy men from the East – and have taken on something of a mythical status. However, they remain obscure enough to be brushed aside by modern scientists.

Michael Werner presents a new type of challenge to skeptics. A fit family man in his 50s, he has a doctorate in Chemistry and is the managing director of a research institute in Switzerland. In this remarkable account he describes how he stopped eating in 2001 and has survived perfectly well without food ever since. In fact, he claims never to have felt better!

Unlike the people who have achieved this feat in the past, he is an ordinary man who lives a full and active life. Michael Werner has an open challenge to all scientists: "Test me using all the scientific monitoring and data you wish!" In fact, he describes one such test here in which he was kept without food in a strictly monitored environment for ten days. Werner also describes in detail how and why he came to give up food, and what his life is like without it. This book features other reports from those who have attempted to follow this way of life, as well as supplementary material on possible scientific explanations of how one could "live on light".[193]

'Michael Werner says:

'My concern is not to persuade people that they should stop eating. My hope is that they will begin to change their way of thinking, including the way they think about eating and drinking.'

'Reviews of the book said:

'[What] can only be ignored with difficulty is the phenomenon itself, for it is crying out to be noticed. One wonders why mainstream science has paid so little attention to it ...[194]

'What Michael Werner wants is to demonstrate that the generally held view of the world being solely physical and solid cannot be the whole story.'[195]

'That's really interesting, Ollie. I wonder why scientists are not more interested in this phenomenon?'

'I think because many avoid studying phenomena that threaten the materialist, 'only-this-dimension-of-matter-is-real', 'life-is-an-accident'

---

[193] Werner, M. and T. Stockli, *Life From Light: Is it Possible to Live without Food? A Scientist Reports on His Experiences* (Clairview Books, 2007). Product description, Amazon.

[194] Harald Walach, Research Professor of Psychology, University of Northampton, and Director of the European Office of the Samueli Institute – back cover.

[195] Neue Luzerner Zeitung – back cover.

paradigm either because (a) it threatens the Darwinian–materialist, 'Life-is-nothing-'super'-Natural-it's-just-an-accident' world view or (b) because *they dare not study* such phenomena for fear of ridicule and serious career damage from the methodologically–materialist thought police and the matterocracy that dominate science and biology today. As one of the reviewers of *Randi's Prize* said:

> '[Psychic and 'super' Natural phenomena] have been studied in painstaking detail for 150 years by highly qualified scientists, including, in recent decades, *some of the most carefully executed scientific experiments ever conducted* with multi-layered experimental controls that put other fields of science to shame. Because researchers in this field are under unrelenting, often vicious assault, they control even against absurdly improbable and unrealistic forms of cheating and fraud among other things, problems that most scientists don't have to think about at all (imagine trying to work in that environment). Statistically and taken as a vast body of work, their results are rock solid . . .'

'I see what you mean. It's not easy to go against the consensus. After all, Galileo hesitated for many years before publicly showing his support for Copernicus' heliocentric system for fear of the ridicule of his academic colleagues.'

'Exactly. Yet, as Bruce Greyson says:

> "*it is the scientific method of empirical hypothesis testing*, rather than a materialistic philosophy, that has been responsible for the success of science in explaining the world. *If it comes to a choice between empirical method and the materialistic worldview*, the true scientist will choose the former"'[196]

'The trouble is, the whole point, Ali, of the materialist philosophy that has dominated science and biology, particularly since Darwin, is the **non-negotiable belief** that there are no intelligently causal or higher Natural elements to reality: 'god is dead, Nature, including the Living World, is 'meaningless', it's 'unintelligent' and it's an 'accident', it's an MUA – it's not an 'amazing, ultimate miracle' or an AUM. It may '*look*' amazingly clever and purposeful,' is the thinking, 'but actually it's all utterly, second by second, 24/7 unintelligent and accidental!'

---

[196] Greyson, B. "Commentary on 'Psychophysiological and Cultural correlates Undermining a Survivalist Interpretation of Near-Death Experiences.'" *Journal of Near Death Studies* 26, no. 2 (Winter 2007); p. 142.

'But what about the trillions of stunningly complex, Life–Force driven, digitally CODED, biochemical operations happening in our bodies right now, every second? What about *our* consciousness and sentiency and *our intelligence,* Ollie? How could awareness, subjectivity, minds and *intelligence,* which are all Soul and spiritual qualities, not material ones, 'mindlessly', 'unintelligently' and 'accidentally' arise out of matter that is itself supposedly utterly mindless, unintelligent and 'dead.'

'I agree. These are the contradictions in materialist thought. Thankfully, they can be resolved at a stroke by accepting that Pythagoras, Plato, Aristotle, Aquinas, Bacon, Galileo and almost all the greatest scientists and thinkers who have ever lived's *rational inferences* to a multi-dimensional, multi-vibrational Cosmos of intelligence and purpose, one that includes stunningly intelligent Life and Soul Forces as Living Principles, is correct. A Cosmos where:

> *'consciousness is fundamental,* ... the material universe is derivative from consciousness, not consciousness from the material universe ...'[197]

'As Antony Peake says:

> 'Self-awareness within a seemingly unaware universe simply does not make sense ... [science may be coming to] a model in which mind creates matter rather than matter creates mind.'[198]

'This is the classical, Idealist conception of existence, part of the perennial philosophy, as Aldous Huxley called it, surely one that is more consonant with the data of reality than the materialist wish to reduce those data to 'mindlessness, unconsciousness, unintelligence and accident' – all the way up and all the way down, for ever and ever.'

---

[197] *The Observer* – http://en.wikipedia.org/wiki/James_Hopwood_Jeans
[198] Peake, A. *The Out of Body Experience: The History and Science of Astral Travel.* Watkins Publishing. Text in square brackets added.

12 Aristotle's Four Causes: Without them Nothing Makes Sense

Aristotle noted that all things have four causes: material (made of), mechanical (how they work), formal (shaping ideas), final (purposes). Materialist thinking, by contrast, hypothesises that Life and Mind, the entirety of Nature, can be explained in terms of matter and mechanism alone arising from mindless, unintelligent matter in combination with endless accidents. This doesn't work. It is intellectually hopeless. A 'triumph of idealogy over common sense' as Thomas Nagel put it in *Mind and Cosmos: Why the Materialist Neo-Darwinian Conception of Nature is Almost Certainly False*.

'One of the problems in modern thinking, Alisha, is that an almost exclusively materialist thinking science mistakenly believes that its partial descriptions of Nature's mechanisms in terms of Aristotle's first two causes alone, those of matter and mechanism, 'made of' and 'how they work,' are *full explanations* which they, most emphatically, are not. Full explanations would need to include their formal and final causes, the ideas and purposes behind them.

'Materialist thought does not see that its explanations relating solely to 'matter' and 'mechanism' are only partial explanations. The materialist hypothesis that there are are no intelligent Formal and Final Causes in Nature leads to today's intellectual dead end *where mind denies mind, intelligence denies intelligence and* where some of the brightest minds around are reduced to insisting that we are "nothing but a pack of neurons"[199] and "We're all zombies. Nobody is conscious."[200] It also leads, conceptually if not in reality, to a totally de-intelligenced and de-purposed Cosmic Mystery of Existence.

'This is why it is so important to understand the difference between science (1) cleverly discovering how things 'work' and what they are 'made of' (their Aristotelian mechanical and material causes) which modern science is very good at and (2) how they *arose* in the first place, how they arise right now, in real time, brilliant nanosecond by brilliant nanosecond, day after day after day for millions of years, their intelligent shaping, Life and Soul forces and what they are *for*,

---

[199] Crick F. 1995. *The Astonishing Hypothesis*. Scribner; p. 3.
[200] Dennett D. 1992. *Consciousness Explained*. Back Bay Books, p. 406. Quoted by L. Dossey: 2010. *Is the Universe Merely A Statistical Accident? The Blog. Huffington Post*.

what Aristotle called their formal and final causes, the ideas and purposes behind them. This, modern science is not good at, and *cannot* be good at, due to its core materialistic assumption, which it is reluctant to relinquish or to allow to be tested or falsified, that Nature is a 'no-intelligence-needed-accident' and its core belief that none of Nature's mechanisms are 'for' anything. This is because, in materialism only Life-Forceless, Soulless mindlessness and accident are truly 'real'. A topsy turby thought scheme that has clouded western thinking ever since Darwin and his attempts to explain very Life itself as an unintelligent and wholly accidental phenomenon.'

'But surely science doesn't *have* to be materialist in outlook?'

'No, it doesn't. As Lewontin said:

'It is *not* that the methods and institutions of science somehow compel us to accept a material explanation of the phenomenal world, but, on the contrary, that *we are forced by* our *a priori adherence to material causes* to create an apparatus of investigation and a set of concepts that produce material explanations ... no matter how counterintuitive, no matter how mystifying to the uninitiated. Moreover, that **materialism is absolute,** for we cannot allow a Divine foot, [or *intelligent* causation], in the door.'[201]

'Contrast Lewontin's conception of science as materialism with:

'Science is **a *methodological process of discovering truths about reality*.** Insofar as science is an objective process of discovery, it is, and must be, metaphysically neutral. Insofar as science is not metaphysically neutral, but instead weds itself to a particular metaphysical theory, such as materialism, **it cannot be an objective process for discovery.** There is much confusion on this point, because **many people equate science with materialist metaphysics,** and phenomena that fall outside the scope of such metaphysics, and hence cannot be explained in physical terms, are called 'unscientific'. This is a most unfortunate usage of the term. For if souls and spirits are in fact a part of reality, and science is conceived epistemologically as a system of investigation of reality, then there is no reason why science cannot devise appropriate methods to investigate souls and spirits'[202] [as, in fact, a

---

[201] Richard Lewontin (1997) Billions and billions of demons (review of *The Demon-Haunted World: Science as a Candle in the Dark* by Carl Sagan, 1997). *The New York Review*, January 9, p. 31. Emphasis and text in square brackets added.

[202] Grossman, N. 2002. *Who's Afraid of Life After Death?* p. 10–12.

courageous minority of scientists has been doing for more than 150 years now].

'The problem is, Alisha, that so many people today do equate science with the world view of materialism rather than seeing it as the principled exploration of Holistic Nature consisting in (1) 'ordinary' Chemical Nature and outer Space and (2) sentient, intelligent 'super' Nature, psychic and inner, multi–dimensional, spiritual Space. The motives for doing any kind of science are intelligent interest, curiosity, love of knowledge and joy in discovery which are all *immaterial* qualities of Mind and Soul, not qualities of the insensible atoms of the materialist world view which couldn't care less about such things. If that was all we were it would mean that we really would all be "unconscious" and "zombies", as Dennett put it, and incapable of doing anything, let alone science.

'Materialist thought is, for its own mysterious reasons, implacably hostile to the true Life, Consciousness, Mind, intelligence, Soul and meaning in all things, not just in Humans and Animals but also to the Soul in amazing, sacred, Living Gaia – Gaia Mundi, Anima Mundi. For materialism Gaia is not 'alive', enSouled and enSpirited, as all the world's earlier peoples knew, some still know, she's just a rock – 'dead', insentient, numb, dumb, solid, stolid matter. And all her lakes and rivers and meadows and mountains and plains and oceans and forests too. She's not our immediate, Living, intelligent Source, a vast organic Consciousness in her own right, literally our Earthly Mother in a Soulful, multi-dimensional, multi-vibrational Cosmos of Intelligence and Meaning – Pachamama who needs care and attention and acknowledgement from us just as She cares for us. No, She's just a random rock. Never mind that, like all Living Organisms, She has many amazing self-regulating and homeostatic features that make Life possible, sustaining it for billions of years. Rather, materialist thinking science regards her as just another random effect. Just another meaningless thing in its impossible and unreasonable pantheon of mindlessness, unintelligence and accident.

'Look, just as desert islanders faced with an unknown Rolls-Royce that they have never seen before can study it in complete ignorance of what it's *for, who* intelligently made it and *intelligently* evolved it (or *why),* today's brilliant biologists can study the Life Forms, what they're 'made of' and how they 'work', their first two Aristotelian causes, those of matter and mechanism, without needing to know what their intelligent sources are or the purposes behind them, their last two Aristotelian causes, formal and final. However, having done that, they have *not* proved that Living Nature is a billions year long running *'accidental chemical reaction!'* All they've done is brilliantly discovered how the Life Forms (or the Rolls-Royce) *'work'* and what they're *'made of"*.

'Which is, as you said, but a *half-description,* solely one of matter and mechanism, not idea and purpose.'

'Yes, exactly.'

We continued on quietly for a while, then Alisha said,

'I'm still baffled as to how materialist thinking came to dominate science and biology so heavily? After all there is *no evidence* that 'accidents' ever build 'machinery' of any kind, let alone the complex codes and software required to drive it, but only randomly evolved landscapes which, while complex, lack *specificity* and crystals which, while specific, lack *complexity*.'

'I think, Ali, it's due, above all, to a few simple but very powerful reasons: (1) People realized that the western, religious six day creation story was not literally true, entirely forgetting the understanding that it was a *metaphor* and was never intended to be treated as literally true in a physical–science sense. (2) Others were fed up with the ultra-dogmatic nature of the western religious traditions with their monopolistic claims to spiritual truth; this caused some people to reject religion and spirituality altogether. (3) Nature is harsh and Life can be very cruel (4) Life's intelligent causes cannot be physically seen but only rationally inferred, psychically intuited or spiritually cognized.

'However, as Michael Denton said:

'between the living and nonliving world ... there is *a chasm* as vast and *absolute* as it is possible to conceive.'[203]

'A chasm that 'accidents' could never bridge?'

'Yes, that's the point. The researches and calculations of thinkers like Hoyle, Dembski, Gitt and Meyer have shown as a matter of mathematical and statistical fact that Life could never have arisen by random and unintelligent processes. That's not merely highly improbable, as all have always agreed, it's impossible. Life and Mind are more than chemistry. And we have no need to look for random events to bridge that chasm if we agree to the classical proposition that Life on Earth, in its Living, spiritual essence, is neither mindless, unintelligent nor accidental – which is not to say that random events play no part in the dramas of existence but that they are not its most central and essential causes.

---

[203] Denton, M. 1985. *Evolution: A Theory in Crisis*. Burnett Books, p. 249. Emphasis added.

'Classical physical–science is based on measurements, calculations, analysis and, when circumstances permit, repeatable experiments using the physical senses or extensions of them with the intellect brought to bear on the results at all stages. However, when it is taken hostage by exclusively materialist thinking it is **radically prevented** from acknowledging Aristotle's last two causes – formal and final, shaping ideas and purposes – anywhere in Nature, because they imply *purpose* and *intelligence*.

'Hence the materialist opposition to the modern theories of intelligent design which merely confirm Aristotle's classical insight that the Natural world is *pervaded* by intelligence and purpose. This should be an intellectual relief. We are intelligent and we emerge from a multi-vibrational, multi-frequency matrix, mysterious as it is, that, also, is intelligent. Blue is back to being blue and is not green anymore. Nature *really is* as clever and teleological as she looks. We do, after all, live in a multi-dimensional cosmos of purpose and meaning.

'By contrast with modern physical–science, spiritual–science, of which the world's religions, spiritual, esoteric and psychical research traditions are all parts, has, by definition, always acknowledged Aristotle's last two causes, ideas and purposes, and the multi-dimensionality, intelligence and teleology inherent in Nature. Spiritual–science is based not only on the data of outer, physical reality but also on information and evidence obtained by inner vision and extrasensory perception, faculties latent in all of us that we can all develop if we wish[204] The trans-dimensional data collected by psychical and extra–sensory observations can then be subjected to intellectual scrutiny by both the original researcher and by others with or without the developed extrasensory abilities to collect the primary data in the first place.

'For example, at the end of the nineteenth century and early in the twentieth the theosophists, Annie Besant and Charles Leadbeater, trained in yogic, spiritual–science techniques, explored the sub-atomic world using trained, inner vision. In *The Secret Life of Nature,* Peter Tompkins describes his investigation to see if there "really was an acceptable correspondence between the theosophists description of material atoms and the 'reality' of orthodox physicists". He says:

'To find out I went in search of the first qualified theoretical physicist to reevaluate the theosophists' pioneering work in

---

[204] Steiner, R. 1918. *Knowledge of Higher Worlds and its Attainment.* Rudolf Steiner Press. 6th Edn. 2004; provides exercises for higher–sensory development. Courses at the College of Psychic Studies, London also provide such opportunities.

Occult Chemistry,[205] Dr Stephen M. Phillips, a professor of particle physics. Phillips's book *Extrasensory Perception of Quarks*, published in 1980, while dealing with the most advanced nuclear theories, including the nature of quarks, postulated particles even smaller than quarks as yet undiscovered by science. Analyzing twenty-two diagrams of the hundred or so chemical atoms described in *Occult Chemistry* ... at the turn of the century, Phillips found it hard to avoid the conclusion that "Besant and Leadbeater did truly observe quarks using ESP some 70 years before physicists proposed their existence". What is more, their diagrams indicated ultimate physical particles even smaller than quarks.

By the time I discovered Phillips on the southern coast of England ... he had checked another eighty-four of the theosophists' atoms: all were seen by him to be 100 percent consonant with the most recent findings of particle physicists. Every one of the 3,546 subquarks counted by Leadbeater in the element of gold could be correctly accounted for by Phillips. Were Phillips's conclusions to be substantiated by his peers, it would adduce evidence that the theosophists with their yogi powers had effectively opened a window from the world of matter into the world of spirit.

Prompt and committed approval of Phillips's conclusions had already come from the noted biochemist and fellow of the Royal Society, E. Lester Smith, discoverer of vitamin B-12. At home in both the mathematical language of physics and the arcane language of theosophy, Smith spelled out his support in a small volume, *Occult Chemistry Re-evaluated* [1982]. And Professor Brian Josephson of Cambridge University, a Nobel Prize winner in physics, was sufficiently impressed by Phillips's radical thesis to invite him to lecture on the subject at the famous Cavendish laboratory in 1985.

Yet few in the ranks of orthodoxy had the courage to risk their positions by supporting anything so wild as the notion that psychics could see better into the basic constituents of matter than could physicists armed with billion-dollar supercolliders.'[206]

'That's very interesting.'
'Yes, and it's incredibly important.'
'Why?'

---

[205] Besant, A. Leadbeater, C. W. & C. Jinarajadasa. 1908. *Occult Chemistry*. Theosophical Publishing House.
[206] Tompkins, P. 1997. *The Secret Life of Nature*, Thorsons, pp. 68–69.

'Because modern science depends, a great deal, on empiricism, on direct evidence, sensory evidence. The philosophical materialism, that dominates science today, hypothesizes as one of its foundational, but *testable and falsifiable,* claims, that the *only senses* that exist are the physical senses. If, however, there are other human senses, non physical senses, which enable us to gain reliable and systematic knowledge about 'ordinary' Natural *or* 'super' Natural Reality, then the materialist paradigm is refuted. Leadbeater and Besant used such other senses, and, therefore, *they falsified the materialist paradigm.* The implications of their work are epoch making.'

'Okay, so given the importance of their discoveries and that Leadbeater and Besant used inner, psychic perception to accurately describe sub-atomic particles before scientists even knew of their existence why isn't that more widely known?'

'Because when their book, *Occult Chemistry,* came out just over a hundred years ago, none of the conventional scientists had ever seen inside an atom. They could not evaluate the theosophists' amazingly accurate descriptions. There was no way to cross check them and it wasn't until a professor of particle physics, Dr. Stephen Phillips, came across a copy of *Occult Chemistry* that their accounts of the particles could be verified. Peter Tompkins says that:

> 'Phillips counted the quarks, and saw that the Theosophists had the correct number of quarks in every element, and the last quark was only discovered in 1997 – the proof of it is self-evident ...
>
> '*The excuses for disbelieving the claims of the psychics are irrelevant in the context of their highly evidential descriptions of subatomic particles published in 1908,* two years before Rutherford's experiments confirmed the nuclear model of the atom, five years before Bohr presented his theory of the hydrogen atom, 24 years before Chadwick discovered the neutron and Heisenberg proposed that it is a constituent of the atomic nuclei, 56 years before Gell-Mann and Zweig theorized about quarks. Their observations are still being confirmed by discoveries of science many years later.'

'Professor Phillips himself published a book, *Extra-Sensory Perception of Quarks*[207] confirming his findings but, according to Peter Tompkins, although it was a very interesting book it was:

---

[207] Philips, S. M. 1980. *Extra-Sensory Perception of Quarks.* Theosophical Publishing.

'too scholarly – too much mathematics and physics in it for the general public. For scientists themselves, it shatters their whole premise and *if they don't want to look at something, they **won't**.* In fact they will do their best to put you down …'

'Tompkins was asked whether the fact that modern quantum physics has confirmed the clairvoyants' descriptions of sub-atomic matter mean that their extra–sensory descriptions of the spiritual hierarchies, the worlds of the angels, nature spirits, devas, fairies, sylphs, undines and so on could also be taken as real? He replied:

'Absolutely. If a hundred years ago the Theosophists were absolutely correct in the number of quarks which they described, then one has perforce to look at their detailed descriptions of nature spirits. Obviously they are using effective clairvoyance. When you find that the shamans all throughout South America, for instance, describe the same phenomena as the shamans in the Far East, rationally you must take it into consideration as being possibly real. Why would people accept the physicists' description of an atom, which the general public cannot see, and not accept the description of the nature spirits that they cannot see? They accept the religious notions of spirit, though they cannot see them, why not accept the possibility of nature spirits and angelic hierarchy? …

'The whole Christian persecution of witchcraft, and the [denial of the] whole world of nature spirits, is what has alienated us from a healthy planet. It's only when we get back in touch and accept their presence that the whole thing falls back into place and we take our place in the cosmos. Otherwise we're totally alienated from a healthy planet and from our role in the cosmos.'[208]

'That is all extremely interesting, Ollie.'

'I agree, it is. It clearly shows that we can know and explore the Cosmos by inner, mental means and that, as the world's yogis and shamen have always known, reality is alive, intelligent and accessible in ways that our current materialistically oriented science just doesn't want to know about. Our western religious tradition also doesn't want to know about this, which, ironically, is a gift to the de-Souling, de-intelligencing forces of materialism. Yet despite the materialism of our science and a mainstream western religious tradition that also seems to regard Nature as Soulless,

---

[208] *Kindred Spirit Magazine.* June, 1998.

the knowledge that the Cosmos is multi-dimensional, intelligent, alive and singing cannot be kept at bay for ever. The spiritual–scientist yogi–adepts have always known that the Universe is an emanation of Consciousness and Intelligence, Spirit and Soul before it is a place of 'matter', and one that is not, fundamentally, 'material' at all.[209] Rather it is 'made' out of high vibrating, intelligent, living forces and energies, which, ultimately disappear into quantum e—m—p—t—i—n—e—s—s, a Cheshire Cat's Smile of a Divine Dream that the yogis call Maya. Mind and Being are eternal but 'matter' and 'materiality' are temporary.'

'A well known body dedicated to the furtherance of modern spiritual–science is the College of Psychic Studies in Kensington, London. The College was founded in 1884 by a small group of distinguished people engaged in psychical research, some notable in science, others Anglican clergy, who had the courage to go against both materialist and orthodox religious convention and sought to show, in a world of increasing materialism, that human personality survives bodily death and that this fact is capable of demonstration.

'This was psychical– and spiritual–science work that should have been applauded and supported by the western religions but generally it was not because it challenged some of their dogmas – such as those relating to 'heaven' and 'hell' and so on. Their failure to endorse this important work was a gift to the de-intelligencing forces of materialism which sought to cut humanity off from the worlds of Soul and meaning and the multi–dimensions of total Holistic Reality altogether with their strange – and now shown to be impossible – hypothesis that the Living World is simply a Soulless *'accident-chemical-reaction'* of billions of years duration.

'In an attempt to counter the materialism of the age the purpose of the College of Psychic Studies was to encourage formal investigation into the psychic and mediumistic phenomena in which many people had become interested at that time. Since then the work has evolved to incorporate the more general exploration of consciousness beyond the body while maintaining the commitment to the development and understanding of mediumship, psychic ability, and healing.

---

[209] Capra, F. 1992. *The Tao of Physics*. Flamingo. 'The … classic exploration of the connections between Eastern mysticism and modern physics. An international bestseller, the book's central thesis, that the mystical traditions of the East constitute a coherent philosophical framework within which the most advanced Western theories of the physical world can be accommodated, has not only withstood the test of time but is ever more emphatically endorsed by ongoing experimentation and research.' Back cover.

'Bring physical–science and spiritual–science together, and we begin to have a more complete understanding of the two sides of Nature: (a) outer, physical Nature, consisting in 'matter' and 'mechanism', from atoms, to molecules, cells, plants, animals, people, planets, stars and galaxies and (b) intelligent, multi-dimensional, spiritual Nature which is an expression of intelligent Beingness, of Spirit and Soul, hologramatically self-fragmented into trillions of lesser sparks, diamonds and flames of consciousness and existence, expressing as all the divine Ideas and Soul purposes seen in outer, visible Nature, and consisting in the subjective dimensions, intelligences and agencies from the tiniest Living Soul and Spirit Beings, the Soul Beings of the Plants and Animals, to Humans, to Angels, Archangels and the intelligent multi-dimensions beyond.'

'So you're saying, Ollie, that physical–science can tell us a lot about the workings of outer, 'ordinary' chemical Nature, of Aristotle's first two causes, those of matter and mechanism, but very little, currently, about Aristotle's last two causes, formal and final, the subjective purposes and meanings of the phenomena of the mineral, vegetable, animal and human kingdoms. Whereas, the world's religious, psychical research, spiritual and esoteric traditions, like theosophy, anthroposophy and the yogic teachings, which are all part of an emerging spiritual–science can, at times, supply us with some useful ideas or insights into these things?'

'Yes. For example, even the simplistic stories in the world's scriptures along the lines of "Supreme Being said, *"Let the Worlds exist!"* and they did," reflect humanity's timeless intuition that Living Nature and the Universe beyond are *not* mindless and accidental, they give us our Formal and Final Causes. They don't, however, explain the second by second mechanics of Life's workings, both now, and across deep time.'

'So, Ollie, it's the job of physical–science working in combination with an up to date spiritual–science to try to find out.'

'Yes, precisely. You see, about 400 years ago Sir Francis Bacon set out an epoch defining program for the development of modern physical–science. One of his insights was that for physical–science to really get going, scientists should stop worrying, in a medieval kind of way, Soulful as it was, about the *ultimate* meanings and purposes of things, their Aristotelian formal and final causes, the *ideas and purposes* behind them, but simply focus, for the time being, on what they were 'made of' and how they 'worked' – that is on matter and mechanism. Later, they could always go back, if they wished, to focusing on and exploring the formal and final causes behind Nature's outer mechanisms. Bacon's program was an outstanding success. It is partly thanks to his insights that the scientific revolution took off and we have all the amazing gadgets and utilities that we enjoy today.

'We are now at the exciting point that it is time for science to return to the exploration of formal and final causes, the meaning, intelligence and purposes of things. We cannot do this by engaging our cool, detached intellects alone but we also have to bring our Souls, hearts and feelings into the equation. We are beings not merely of physical sensing and dry thought but also of intuition and higher–sensory capacities that permit alternative ways of data gathering and making sense of reality. Ways of knowledge gathering that materialism strenuously denies in its mephistophelian wish to cut us off forever from the Divine and the multi-dimensional worlds of true Life, Soul, Spirit and Meaning. Western science also now needs to come to terms with the yogic, currently esoteric, knowledge that our physical body is but one of several bodies or sheaths that we, as Souls, make use of. On death we shed the physical body but continue on in our emotional and mental bodies, as described, for example in the fascinating book, *Your Life After Death* (2011) by Joseph and his amanuensis Michael Reccia.

'Bacon himself, who was a scientific–Idealist, intelligent design thinker, did not for a moment adopt the modern materialist ('accidents-build-clever-machines') attitude that Nature's stunningly smart Bio-Logical (Life and Soul Logical), 'bio-machines' actually did not *have* any intelligent formal or final causes or ideas and purposes behind them – that they were 'accidents'. He would have seen any such idea as irrational as would almost all the world's greatest thinkers and scientists, before and since – Plato, Aristotle, Copernicus, Galileo, Kepler, Boyle, Newton, Leibniz, Faraday, Maxwell, Planck, Heisenberg among many others.

'Eventually, however, classical physical–science, always predominantly idealist in its thinking, was captured by the deadening hand of materialism and began to argue that it could explain Nature's countless, Soul-full, Life-Force-full, 24/7 mechanisms, the Living Cells, the Lions, Frogs, Beetles, Butterflies, Chestnut Trees and People *without any reference at all* to Aristotle's Formal and Final causes, the shaping ideas and purposes involved – the clever, higher-level Natural laws, Life and Soul forces and other processes and agencies behind them.'

'Which would have seemed unreasonable to the founding fathers of science, Aristotle, Bacon, Galileo, Newton, etc?'

'Yes, because they were all Idealists, 'mind-before-mechanism' thinkers. We can explain how the clever Soul–Life–Form Being's outer mechanisms, such as the bio-chemical bodies of frogs or butterflies or sea anemones *'work'* and what they are *'made of'*, just as the islanders can explain how Rolls-Royces work and what they are made of, without knowing, or needing to know, as Bacon advised, what the subtle, superordinate laws, forces or intelligences behind them were and are, right now, second by second. However, that *half-explanation,* useful as it is for practical purposes, is as far as we can go unless we are prepared to *concede* and *rationally infer* that such subtle Higher Natural laws, Life Forces and Soul processes *really do exist* – their intelligent formal and final causes. This, prior to Darwin, is what all the world's greatest scientists and classical thinkers always did. It was, among other things, quite clear to them that something causes the heart to beat and the body to breathe that is more than Soulless, Life-Forceless, mineral chemistry – let alone the non-evidence-based materialist idea of 'accidental' chemistry.

'Today, physical–science is beginning to divide into two hypotheses of reality. Firstly, the materialist, unintelligent design hypothesis that mindless matter + an infinity of accidents explains all. Secondly, the idealist, intelligent design view that all four of Aristotle's causes are necessary for complete explanations of outer Nature, and that *pre-existing* Consciousness and Intelligence combined synergistically, inter–dimensionally and holistically with matter, better describe reality.

'Traditional unintelligent design or UD thinking is great for explaining (relatively) *unintelligent* and randomly evolved phenomena like river valleys but, obviously, cannot be fully explanatory when studying phenomena of intelligent origin. ID or intelligent design theory is, on the other hand, useful for working out when *intelligence* and *intelligent* agencies are involved in the origination and evolution of various phenomena. For example ID theory is used all the time in archeology, forensic science, cryptography and SETI, the search for extra-terrestrial intelligence.

It is also used in biology and natural history now by scientists who apply it to detect the signatures of *intelligent* rather than accidental causation and evolution in the Living World. The reasoned and principled arguments for intelligent design in the Living World are cogently explored in books like *Darwin's Black Box* and *The Edge of Evolution* by Michael Behe, *The Design Inference* by William Dembski and *Signature in the Cell* by Stephen Meyer.

'Mainstream science and biology, however, continue for the time being with the now shown to be impossible hypothesis that Living Nature is just a massive Soulless and Life Forceless chemical 'accident'. It is the rule of methodologically–materialist science that 'accidental functionalities' (an oxymoron), 'purposeless machinery' (another oxymoron) and trillions of 'random events' are to be considered supremely adequate to fill all knowledge–gaps, to explain *all* Natural phenomena, from atoms to galaxies, from particles to people, the Lions and the Butterflies, the Buttercups and the Bluebells, the Eagles wing and the Dolphin's fin, Shakespeare's poetry and Bach's music, Newton's calculus and Einstein's general theory of Relativity.

'You mean, Ollie, that methodologically-materialist or the question-beggingly named 'naturalist' thinking in science and biology is, fundamentally, 'accident-of-the-gaps' thinking? That whatever we don't currently understand we fill our knowledge–gap with words like 'random', 'mindless', 'unintelligent' and 'accident'? But where's the evidence that 'mindlessness', 'unintelligence' and 'accidents' are 'creative' or 'evolutionary' agencies? No machinery we make ever arises by us randomly throwing the parts around.'

'You're right, the primary evidence for the materialist world view that has such a grip on modern scientific thinking is the physical invisibility of Life's subtle, intelligent causes whose logical and necessary existence the materialist view point refuses to *rationally infer* or to take any notice of collective humanity's intuitions and psychic and spiritual cognitions that they are real.

'What many people don't realize is that the materialist and Darwinian thinking, which so dominate science today, are all about passionate attempts to find no-*intelligence*-needed explanations for Existence in general and for the Living World in particular. Partly, they are theodicies, ways to try to get around the problem of what the Victorians called 'natural evil'. How could a 'good' God or benevolent Higher Nature, they reasoned, be responsible for all Nature's harshness and cruelty?'

'They had a point there, though, didn't they, Oliver?'

'Look, as I said before, I'm not arguing for Nature all 'perfect, good and kind' but for Nature's *intelligence* and purposefulness. The intelligence and meaningfulness of existence are not nullified because aspects of it do seem so harsh. Our capacity to even notice the harshness and to care about it is a mental, Soul and emotional capacity – it is not a capacity of the kind of insentient, insensible 'matter' of materialist faith and belief. Aside from the problem of natural evil Darwin and his contemporaries further concluded that:

> "Because we cannot directly see Life's 'super' or higher Natural causes and because the western, religious accounts of creation are so simplistic, perhaps it makes more sense to think of existence as 'mindless, unintelligent and accidental'. Now let's use our bril-

liantly clever minds and intelligences to do science and to work out how all these purposeless and random mechanisms from astounding atoms to vast galaxies, from fabulous gravity to amazing light, magical magnetism and exciting electricity, from digital DNA CODING, "more advanced than any software", to Living Cells, "more complex than Jumbo Jets", from Ants to Elephants, from Buttercups to Sweet Chestnuts, from Bacteria to People all 'unintelligently' and 'accidentally' arise and meaninglessly and purposelessly work.'

'In other words, Alisha, modern science and biology are now purposefully dedicated to the task of seeking to *intelligently* explain how utterly *unintelligent* and purposeless or non-teleological Nature is and how meaninglessly and *unintelligently* it gives rise to the brilliant minds and *intelligences* of the modern scientists and philosophers who can tell us how mindless and unintelligent it all is based on nothing but the fact that the world's scriptures are not biochemistry textbooks and that total Reality's intelligently causal, but *non-physical,* laws, forces, and processes cannot be seen in *physical* telescopes and microscopes but can only be intuited or rationally inferred – neither of which things materialist thinking science and biology are keen to do.

'This is quite a reversal of the original scientific enterprise founded by the ancient Greeks, like Pythagoras, Plato and Aristotle which, centuries later, was taken forward by Bacon, Galileo, Kepler, Newton, Leibniz and Boyle etc who all pursued science as an exercise in the study of Nature's *intelligence* and intelligent design because they believed the Cosmos to be a rational place, in many if not all respects, and, therefore, due to its *inherent intelligence,* purposefulness or teleology it would, they reasoned, be capable of being described and explained intelligently by the intelligent and teleological activity that we call the practice of Science – which is, precisely, how it turned out.

'But why, Ollie, the so very passionate resistance in modern, mainstream science and in biology to the very idea that existence might be *intelligent* and meaningful, an *intelligent* Cosmic Mystery rather than a *meaningless* Cosmic Mystery and the so very vehement opposition to the idea that scientists are perfectly capable of telling the difference between random and relatively *unintelligent* phenomena in Nature (like randomly evolved river valleys) and *intelligent* and purposeful phenomena in Nature (like intelligently evolved Life Forms)?'

'I think the reasons are psychological, not evidential as, whatever Life, Mind and sentiency all precisely are, trying to explain them all as an expression of 24/7 mindlessness, unintelligence and accident is absurd, especially as Cellular Life, it turned out, is orders of magnitude more intelligently complex than Darwin and his contemporaries imagined it to be.

'Other *psychological* reasons for trying to explain the magical and Soulful Living World of the Plants, Animals and People as a never ending *'accidental chemical reaction'* are sadness at the harshnesses of Life, disappointment at the physical invisibility and intractability of the intelligently causal multi-dimensions, vibrations and frequencies of total Holistic Natural Reality and frustrations with the organized religions that have let people down in various ways.

'Contributing to the rise of materialist thinking were western organized religion's lack of interest in modern spiritual–scientific research, for example the insights of theosophy and anthroposophy,[210] the discoveries of spiritualism and psychic research and the accompanying confirmations of the reality of Life after death.[211] These modern spiritual–science findings are so much more sophisticated than the simplistic, medieval notions of the divine that the church taught for so long and which made less and less sense to many modern thinking people and caused them to turn away from religion and spirituality altogether.

'Historically, there were also times when the church held back the development of physical–scientific understanding and persecuted those who had alternative ideas. This, ironically, is a role now occupied by the new materialist high priesthood in science which oppresses and launches witch-hunts against modes of thought that do not conform to its dogmatic 'only-matter-matters' paradigm. For example the materialist attacks on complementary modes of healing and therapy – including herbal, homeopathic, crystal, spiritual and energy healing, and, of course, the modern theories of *intelligent* evolution and intelligent design.

'However, despite what some people claim, science does not equal materialism. Science simply describes the systematic and principled attempt to acquire knowledge of what is *actually true,* be it concern-

---

[210] Described and founded by Helena Blavatsky and Rudolf Steiner respectively.
[211] Fontana, D. 2005. *Is There an Afterlife?: A Comprehensive Overview of the Evidence.* Schwartz, G. E. Phd. & W. L. Simon. 2002. *The Afterlife Experiments: Breakthrough Scientific Evidence of Life After Death.* Carter, C. 2010. *Science and the Near Death Experience.*

ing physical phenomena and the exploration of matter and outer space or non–physical, spiritual phenomena, involving the exploration of consciousness, inner space and the multi-dimensions such as in the modern fields of OBE and NDE research[212] and the research into reincarnation[213] carried out in focused, systematic and principled ways. Some scientists are materialist ('accidents-build-and-evolve-clever-mechanisms') thinkers and some are philosophically idealist ('intelligence-builds-and-evolves-clever-mechanisms') thinkers, but what is actually true is not a matter of world view it's a matter of empirical truth. As Bruce Greyson says:

> "*it is the scientific method of empirical hypothesis testing,* rather than a materialistic philosophy, that is been responsible for the success of science in explaining the world. If it comes to a choice between *empirical method* and the materialistic worldview, the *true scientist will choose the former*"[214]

---

[212] For example, NDEs or near death experiences are now formally studied by the International Association for Near Death Studies, founded in 1981. IANDS publishes the *Journal of Near Death Studies,* a peer-reviewed, scholarly journal.

[213] Stevenson, I. 1980. *Twenty Cases Suggestive of Reincarnation.* Second Edition, Revised and Enlarged. University of Virginia Press.

[214] Greyson, B. "Commentary on 'Psychophysiological and Cultural correlates Undermining a Survivalist Interpretation of Near-Death Experiences.'" *Journal of Near Death Studies* 26, no. 2 (Winter 2007), p. 142.

13 Spiritual Science: The Soul Persists – There is Life after Death

The materialist idea (ideas are not material) that Mind, Intelligence and Consciousness are the 'accidental' byproducts of primordial soup and that there are no Life and Soul Forces that in-form and organ-ize bio-chemical bodies when Alive and depart on 'death' is refuted by the modern evidence for Life *after* Life.

'If particle physicists report their findings we believe what they say. Yet, most of us have no way of testing the truth of their claims. The reply to that is: 'Look, they'll have been peer reviewed and if you study hard enough you yourself will be able to access the information directly. That is very different to so-called psychic and spiritual information which you just have to take on trust.'

'This, though, is not strictly true. Just as with particle physics if we have the interest and talent we can also train to be spiritualist mediums or psychics, at bodies like the College for Psychic Studies in Kensington, London and discover for ourselves whether or not there is any truth in this matter. Most of us simply don't have the interest or enthusiasm to do so, just as most of us don't have the enthusiasm or the talent to become particle physicists – but, theoretically, if we had the interest and the ability, we could.

'I was once given some information, at a social gathering, by a psychic–medium, who'd I'd never met before and never since, relating to my deceased grandmother. The information was precise. He said: "Your grandmother married twice and she survived both husbands, she was a spiritual lady, a genuinely good person." Now, the overly skeptical, cynical view is that this person has nothing better to do than to go up to people, he doesn't know, say the same thing each time until he gets lucky! What for? He wanted nothing from me, we only spoke extremely briefly, I didn't give him anything and I never saw him again. Some dismiss such accounts as 'merely' mind reading or telepathy as though that in itself is not a challenge to the purely materialist, 'no 'spooky' action at a distance' world view. I was certainly not thinking about my grandmother at the time.'

'But, that's just an anecdote, Ollie.'

'Yes, you're right. By contrast with my anecdote, much carefully controlled research has been conducted over the last century or so demonstrating that properly trained psychics and mediums can provide genuinely good evidence for survival of the trans-dimensional shift we call 'death', and that, in fact, as far as our conscious selves or individualities are concerned, as opposed to our chemical body 'diving suit', 'there is no death'.[215] An extremely thorough study was that concerning Frederick Myers and the cross-correspondences.

'Frederick W. H. Myers was a Cambridge classics scholar, a brilliant man who advanced a transmission theory of the mind–brain relationship over a century ago, a theory taken up and developed further by his friend and colleague Williams James.[216] Myers was also one of the founders of the British Society for Psychical Research or BSPR. He sought to find a way to prove that information transmitted through mediums could *not* 'merely' be the effects of telepathy between the living but was real evidence of Life after death. He came up with the idea of cross correspondences – a series of messages to different mediums in different parts of the world that on their own would mean nothing but which when put together would make sense. He and his colleagues at the BSPR believed that this would provide convincing evidence of survival.

'After he left the physical, in 1901, more than a dozen different mediums in different countries began receiving a series of incomplete scripts that they channelled by means of automatic–writing. The pieces were signed 'Frederick Myers'. The scripts related to and contained obscure references from the Classics that made no sense on their own. The writings instructed the mediums to contact a central address where the scripts were assembled and were found, finally, to make sense. Later, Myer's colleagues at the British Society for Psychical Research, Professor Henry Sidgwick and Edmund Gurney, also transmitted and signed such scripts after they too had passed on. More than 3000 scripts were transmitted over 30 years. Some of them were more than 40 pages long. Together they fill 24 volumes and 12,000 pages.

'The mediums used by Myers and the other after–Life communicators were not highly educated and, therefore, had little knowledge of the

---

[215] Schwartz, G. E. PhD. & W. L. Simon. 2002. *The Afterlife Experiments: Breakthrough Scientific Evidence of Life After Death*. Atria Books. Fontana, D. 2005. *Is there an Afterlife: A Comprehensive Overview of the Evidence*. O Books. McLuhan, R. 2010. *Randi's Prize: What Sceptics Say About the Paranormal, Why They Are Wrong, and Why It Matters*. Matador.

[216] Kelly, E. F. & E. W. Kelly, A. Crabtree, A. Gauld 2009 *Irreducible Mind: Toward a Psychology for the 21st Century*. Rowman and Littlefield.

Classics. The messages were beyond their levels of learning. On one occasion one of the mediums, Mrs. Coombe-Tennant, was conducting a discussion using automatic–writing between Professor Sidgwick, passed on, and his living colleague G. W. Balfour on the 'mind-body relationship', 'epiphenomenalism' and 'interactionism'. She complained that she had no idea what they were talking about and became annoyed at being expected to transmit such difficult material. Myers himself said it was extremely difficult to transmit his messages from the spirit world to the mediums, that it was like, "... standing behind a sheet of frosted glass which blurs sight and deadens sound, dictating feebly to a reluctant and somewhat obtuse secretary".[217] The original documents are on file, for those with the patience, to study. One such was Colin Brookes-Smith. After conducting his research he stated, in the *Journal of the BSPR*, that survival should now be regarded as sufficiently well-established to be *beyond denial by any reasonable person.*[218]

'Colin Wilson, world famous author of *The Outsider, Superconsciousness, The Occult* and many other books said:

> 'For anyone who is prepared to devote weeks to studying [the Myers cross-correspondences], they prove beyond all reasonable doubt that Myers, Gurney and Sidgwick went on communicating after death.'[219]

'So why has this not been accepted and become mainstream, Ollie?'
'Because the western intellectual world has become overwhelmingly materialist in its outlook and tends to ignore evidence that falsifies the materialist interpretation of reality. As Neil Grossman found, when many "scientists and philosophers are confronted with the evidence, their reaction is often anything but rational."[220] Chris Carter explains that it is the confusing and unjustified conflation of science with the philosophy of materialism that leads to the widespread practice of ignoring evidence that refutes the materialist world view and labeling it as 'unscientific'. Yet:

> 'scientific beliefs are **not** held as dogmas, but **are open to testing and hence possible rejection**. Science cannot be an objective process of discovery if it is wedded to **a metaphysical belief that**

---

[217] Tyrrell, G. N. M. 1938. *Science and Psychical Phenomena*. Methuen & Co. Ltd; p. 297.
[218] Quoted by V. Zammit, victorzammit.com: "The objective evidence for the after Life has nothing to do with religion or personal belief."
[219] Colin Wilson, 1987; p. 179. quoted by Zammit.
[220] Carter, C. 2010. *Science and the Near Death Experience*. Inner Traditions; p. 236.

is accepted without question and that leads to the exclusion of certain lines of evidence on the grounds that these lines of evidence contradict the metaphysical belief.'[221]

'You mean that if our unquestioned 'scientific' belief is that there cannot be telepathic communication between those passed on and those in this dimension we will simply dismiss out of hand all evidence that says otherwise?'

'Yes, exactly. Yet what is not widely known today is that from the mid nineteenth century onwards the question of the survival, or not, of consciousness and individuality beyond 'death' was studied by some very eminent thinkers and scientists. These scientists all started out as open-minded skeptics – not closed-minded ones who simply dismiss all evidence that does not match their preconceptions. On thorough investigation, their skepticism satisfied, these open-minded skeptics concluded that Soul, Life, Mind and Consciousness were *intelligent* mysteries that continued beyond the cessation of bio-chemical functioning that we call 'death'. They realized that 'death' is but a dramatic trans–dimensional shift of frequency and vibration, the Mind and the Soul shedding an outworn skin, and not the end of individual consciousness, no matter how much some thinkers hope it to be.'

'Surely no one 'hopes' that there is no Life after death, Oliver?'

'You say that, Ali, but some thinkers really do just do seem to find it more satisfying to pour scorn[222] on the all the evidence that humanity's intuitions, not to mention rational inferences and logical deductions, that our existences are not the products of 'accidental' chemistry, continuing for millions of years, without rhyme or reason, and that Consciousness, Sentiency, Mind, Love and Intelligence are ontological essences and Soul qualities that can express in what we call physical matter but can also express beyond it, at other vibrations and frequencies of more subtle 'matter' – that is they are multi–dimensionally real, not imaginary.

'Listen to these unequivocal statements that some of the early pioneers into inter–dimensional communication made once they'd satisfied themselves of the evidence, bearing in mind that these

---

[221] Ibid. p. 237, emphasis added.
[222] Carter, C. 2007, 2012. *Science and Psychic Phenomena: The Fall of the House of Skeptics*. Inner Traditions. Carter, C. 2010. *Science and the Near Death Experience*. Inner Traditions. McLuhan, R. 2010. *Randi's Prize: What Sceptics Say About the Paranormal, Why They Are Wrong, and Why It Matters*. Matador.

thinkers were none of them gullible fools but distinguished and skeptical scientific professionals:

'Sir William Barrett F. R. S., 1844–1925. Professor of physics at the Royal College of Science in Dublin stated:

> "I am absolutely convinced of the fact that those who once lived on earth can and do communicate with us. It is hardly possible to convey ... the *strength and cumulative force* of the evidence." [223]

'Sir Oliver Lodge F.R.S., 1851–1940. Professor of physics, University College in Liverpool, England and later principal at the University of Birmingham. World famous for his pioneering work in electricity and radio and for developing the spark plug:

> "Communication is possible. I have proved that the people who communicate are who and what they say they are. The conclusion is that *survival is scientifically proved by scientific investigation.*"

'Professor Camille Flammarion, 1842–1925. World renowned astronomer who founded the French Astronomical Society. He investigated psychic phenomena for more than 50 years. He said:

> "I do not hesitate to affirm my conviction, based on personal examination of the subject, that any man who declares the phenomena to be impossible is one who *speaks without knowing what he is talking about;* and, also that any man accustomed to scientific observation – provided that his mind is not biased by preconceived opinions – may acquire a *radical and absolute certainty* of the reality of the facts alluded to."

'Professor James J. Mapes, 1806–1866, an expert in chemistry, set out in the 1850s to rescue his friends involved with the then spiritual mediumship craze. After investigating many mediums he changed his views and stated that:

> "The manifestations ... are so conclusive [that they establish]: First, that there is a future state of existence, which is but a continuation of

---

[223] This quote, and the others that follow, compiled by Michael E. Tymn, Vice-President, ASPSI, www.aspsi.org; emphasis added. "It wasn't long after the birth of modern Spiritualism in 1848 that scientists and scholars began investigating the phenomena. Many of them started out with the intent of showing that all mediums were charlatans, but one by one they came to believe in the reality of mediumship and related psychic phenomena." M. E. Tymn.

our present state of being … Second, that *the great aim of nature,* as shown through a great variety of spiritual existences is *progression,* extending beyond the limits of this mundane sphere …"

'These eminent scientists and thinkers, Ali, were all skeptics, albeit open–minded ones, they were not 'nuts' or 'fundamentalists' of any kind and it was only after a great deal of investigation that they accepted the reality of medium–channelled communications from those passed on and the evidence for Life after death. Many of them were highly practical people whose major discoveries in other fields fundamentally changed the way humanity works and lives. They were free thinkers and braved opposition not only from materialists but also from traditional church and religious authorities.

'But why would the traditional religious authorities oppose the modern evidence for Life after death? Surely it would do so much to strengthen their case for a multi-dimensional, multi-vibrational Cosmos of Intelligence and meaning rather than the de-intelligenced and de-purposed materialist Cosmos of 'mindlessness' and 'accident'? '

'You'd think so, Ali, but you are being somewhat naive. You see the modern research into Near Death experiences, verified medium channelled communications from those passed on,[224] research into reincarnation and so on all challenges many long held religious dogmas. For example, dogmatic religious ideas about 'heaven' and 'hell'. The modern evidence is that in many respects Life after 'death' is similar to physical life, we still have bodies, made of a more subtle or parallel dimension of matter, minus some of the physical limitations, and that it is our attitudes and mental and emotional states that govern 'where' we find ourselves in those dimensions. In other words 'heaven' and 'hell' are reflective of inner mental and emotional states just as, in many ways, they are in this exceptionally dense dimension. Any after-Life mental and emotional suffering is temporary, not 'eternal', and transforms as changes of heart, mind and attitude take place.

'Another thing that conflicts with many traditional religious teachings is that the message comes through, again and again that it is *not* what you profess to believe, in religious terms, it is not the creed you adhered to, or even whether you had one at all, but how you *were* during physical Life, how you thought and behaved, not your particular theology, or lack of it, that is a measure of your spiritual and Soul progress in any given physical Life time.

---

[224] Schwartz, G. E. Phd. & W. L. Simon. 2002. *The Afterlife Experiments: Breakthrough Scientific Evidence of Life After Death.*

'You see, if traditional religious authorities cannot, to some extent, frighten people with stories of 'hell' and the insistence that they have a monopoly on inter–dimensional and spiritual truths, that their particular religious way is the only way, then they lose a good deal of their traditional control. This is one reason, I think, why they show a marked lack of interest in the modern evidence for and knowledge of Life after 'death'. They don't necessarily wish people to realize that the paths to Soul and spiritual growth and enlightenment depend, above all, not on outer creeds and beliefs, which are but sign posts and tools, but on love, kindness and on going within to be still, to meditate, to visualize and to pray, but tend to insist that it's their way or the highway (to somewhere unpleasant).

'Okay, I see what mean and, now I come to think of it, I think you're being naive too.'

'In what way.'

'Well the christian and other western religions are in an awkward position. Firstly, as you say, they tend to disapprove of inter–dimensional communication because it challenges their claimed monopolies on spiritual truths. But secondly, they have to be cautious about endorsing any modern evidence for the survival of death in case anti–spiritualist, materialist thinkers come along and try to debunk it, which, of course, they do.

'You see, Oliver, if the religions keep to their scriptural authorities and dogmas and say that their views of Life after Life, heaven, hell and so on are all based on faith and ancient evidence there's not much the skeptics can do to challenge it except disagree. But the modern evidence for Life after the trans-dimensional shift we call 'death' because it is contemporary and not based on ancient authority, is a much more serious threat to the materialist paradigm so it is attacked even more fiercely. So the religious authorities observe from the sidelines to see how the battle between the materialist ('mindless-accidents-create-minds-so-there-can-be-no-life-after-death') ideology, currently dominant in science, biology and psychology, and the idealist ('mind-before-minds,' 'intelligence-before-intelligent-mechanisms') philosophy, currently the minority position in science, plays out before showing their own pro-Life-after-death hand.'

'Okay, Ali, there's something in that, but isn't it shortsighted? Firstly, the modern evidence for the continuity of consciousness and for Life after death is currently dismissed out of hand by the materialist world view which dislikes the idea that consciousness is anything but an 'accident' and, therefore, it opposes the modern evidence that consciousness can exist outside the body which, in any case, it refuses to study.

'How many materialist skeptics have ever bothered, for example, to properly evaluate the Myers cross-correspondences, or to read up on Dr. Gary Schwartz's *The After-Life Experiments* in which:

> '*under controlled laboratory conditions.* In stringently monitored experiments, leading mediums attempted to contact dead friends and relatives of 'sitters' *who were masked from view and never spoke, depriving the mediums of any cues.*'[225]

'How many have read David Fontana's scholarly and 500 pages long *Is There an After Life: A Comprehensive Overview of the Evidence.* They don't bother to study these serious pieces of work because they do not seem to be genuinely curious, open minded and scientific minded: that is willing to examine the evidence and to follow it wherever it may lead. Rather, they adhere to a dogmatic materialism which is, for them, like a fundamentalist religion whose tenets can never be questioned. As this reviewer of *Randi's Prize* said:

> 'the [skeptical] opposition is superficial, ill-informed, misleading, *sometimes blatantly dishonest* and occasionally bordering on the hysterical, as if afraid of the implications – which are indeed formidable.'

'And another:

> '[The skeptics] systematically failed to really engage with their subject matter, keeping their distance from doing actual research and even from becoming genuinely acquainted with the original literature. . . . with the sole intention of destroying it . . .'[226]

'The modern evidence for the reality of out of body and near death experiences and for Life after death, that we really are minds and Souls apart from bio-chemical bodies, that consciousness is paramount, needs the support of the religious authorities and the world's spiritual traditions rather than their studied indifference. Surely, it is in the religions' long term best interests to have their ancient claim, that death is not the end, validated – even if the modern evidence

---

[225] Schwartz, G. E. PhD. & W. L. Simon. 2002. *The Afterlife Experiments: Breakthrough Scientific Evidence of Life After Death.* Atria Books.
[226] McLuhan, R. 2010. *Randi's Prize: What Sceptics Say About the Paranormal, Why They Are Wrong, and Why It Matters.* Matador. Amazon reviews by M. R. Barrington and R. Perry.

challenges some of their ancient notions of 'heaven' and 'hell' and so on. Religion needs to integrate modern evidence and understanding of the dimensions 'next door' and to become part of a modern spiritual–science, Ali, not remain based forever on ancient dogmas.

'For example, is Christianity necessarily incompatible with the growing modern evidence for reincarnation and repeated human lives? The concepts of 'rebirth into spirit' or 'salvation' don't disappear, they simply become aligned with the ancient knowledge, never forgotten in the east, that the human Soul goes through many physical life-times as it evolves and matures, developing capacities and talents and also moving gradually from primarily survival based and self-centered ways of being to progressively more all-centered ones. The spiritual goals remain the same: Soul progress, Soul maturity, the garnering of experience, knowledge and ultimately full spiritual enlightenment or salvation.'

'But why is there such a system of repeated Lives, Oliver?'

'Who knows? No one knows why there is a need for *even one* Life. The system is the system. No one even knows *why* there is anything at all, even just one electron, rather than nothing at all and saying that (1) 'Accidents', from the Big Bang on, did and do it, 24/7, as the materialists do, (2) Cosmic scale 'intelligence' did and does it, from the Big Bang on, as the idealists do or (3) God or Supreme Being does it, from the Big Bang on, as religious and spiritually minded people do does not answer that ultimate question.

'The idealists say to the materialists: 'Powerless, impotent unconsciousness, mindlessness, unintelligence and accidents, with no qualities of any kind of Being at all, do not have the potential to generate even one quark, let alone a universe that is 20 billion light years across.

'The materialists reply: 'So where does the universal awareness, intelligence and stupendous, purposeful potency of Being that you idealists and religious people believe in as the Final Source or Sources of All that Is come from? Who made them or It then?' The idealists and religious and spiritual people cannot answer, for no one knows.

'All we can do is try to work out which hypothesis of reality, mindless, unintelligent and 'accidental' non-beingness or blazingly intelligent and purposeful Beingness, makes more sense as the Final Source and Uncaused Cause of all that is.

'Is the materialist concept of preexisting unconsciousness, insentience, impotence, mindlessness, unintelligence and some kind of mindless, unintelligent, accidental-universe-generating-mechanism or MUA–UGM the better explanation?

'So,' the scientific–idealists, the religious and the spiritual people ask the materialists:

'Who made your pseudo-uber-god, the mindless, unintelligent, accidental-universe-generating-mechanism or MUA–UGM you believe in as the mindless, unintelligent and accidental source of all that is?'

'The materialists reply: *'No one did, it just is!'*

'This sounds, however, suspiciously like the answer that the Idealist and religious thinking people give to the materialists when they ask:

'So who made the God, Brahman, Allah, the Buddha Nature, the Tao or the Supreme Intelligent Beingness that you say is the final Uncaused Cause and Ultimate Source of all that is?'

'The idealists and the spiritually minded people reply: *'No one did it just is!'*

'If we agree with the materialist view, then it's the job of science and brilliantly intelligent scientists to work out how the impotent, mindless, unintelligent and accidental or MUA system, we are so meaninglessly, unintelligently and accidentally in, works. Although why it should 'work' at all, being so 'mindless, unintelligent and accidental' is anyone's guess. If we go with the idealist view then it's the job of science, both physical and spiritual science, to work out how the astonishing, intelligent, purposeful system of amazing, ultimate miracles or AUMs we are in, actually works. Perhaps one day we'll even know why it is as it is.

'For now, the things that we know are that all things evolve, either randomly or intelligently. The idea that human consciousness itself is in intelligent evolution, as part of that understanding, makes sense. We also now know, based on modern research, not ancient religious authority or faith, that human consciousness is something, ultimately, that is apart from the body, it is immortal and survives the death of the body. The evidence is in, it is just a fact of the system or matrix we are in – strangely unwelcome as that is to materialist thought. As neuro–psychiatrist Dr. Peter Fenwick and the UK's leading clinical expert on near death experiences or NDEs put it:

'after assessing the evidence, *there can no longer be any doubt that there is life after death.*[227]

---
[227] Fontana, D. Ph.D. 2005. *Is There an Afterlife?: A Comprehensive Overview of the Evidence*. O Books. Amazon review; emphasis added.

'It also seems clear, although the evidence is not yet as well established, that we have repeated human lives involving a process of individual Soul growth, learning and evolution. This expert on near death experiences writes:

> '[Near Death] experiencers state over and over again that the human soul evolves. They speak of cycles ... and that one lifetime is hardly enough to perfect the Self they really are on its journey back to the One True Source of All. ... We have the choice, they say, free will, to lengthen or shorten the process. I have never heard any of them use reincarnation as some sort of lame excuse to avoid the responsibility and the effort needed to develop the life at hand.'[228]

'This is about as much as we can currently say.'
'It's disappointing, Oliver, that we can't go further than that, isn't it?'
'Look, in time, we may be able to. Right now we are still establishing basic principles. Our contemporary, materialist thinking science still doesn't even accept that we exist within an intelligent, multi-dimensional Cosmic system as opposed to a mindless and accidental one. This is why it is so important to get the modern knowledge that Sentiency, Life, Mind and Soul are not Darwinian 'accidents' of 'accidental' bodies, anymore than TV Shows are 'accidents' of 'accidental' TVs, more widely established. This is also why the modern evidence for the trans-dimensional reality of Life after physical 'death' is so important.

'Listen to just a few more comments of the early and more recent pioneers who have seriously investigated telepathically channeled communications and spiritual mediumship demonstrating the reality of Life after Life:

'Dr. Alfred Russell Wallace, 1823–1913. Co-originator with Charles Darwin of the, as it was first known, Darwin–Wallace theory of evolution said, quite dryly:

> "My position is that the phenomena of Spiritualism in their entirety *do not require further confirmation.* They are proved quite as well as facts are proved in other sciences."

'Wallace was, like Darwin, a materialist until he began investigating psychical and spiritual mediumship in the 1860s. He also realized that

---
[228] Atwater, P. M. H. 1994. *Beyond the Light: Near Death Experiences.* Thorsons; p. 116.

his and Darwin's theory could not genuinely explain Life's stunningly complex bio-functionalities and – in its entirely materialist forms which deny the Life Forces and the Soul in all things – washed his intellectual hands of it. Wallace continued to believe in evolution but as an intelligent and purposeful phenomenon, not as the blind and no-intelligence-involved phenomenon that Darwin had in mind. Echoing Wallace's view, Dr Gustave Geley, 1868–1924, Professor of medicine at the University of Lyons said:

"The facts revealed necessitate "the complete overthrow of materialistic physiology... *the materialistic conception of the universe is false...*"

'More recently, Dr Gary Schwartz, Ph.D said:

"I can no longer ignore the data [on Life after Life] and dismiss the words [coming through mediums]. They are as real as the sun, the trees, and our television sets, which seem to pull pictures out of the air."

'Dr. Gary Schwartz, a Harvard Ph.D., served as a professor of psychology and psychiatry at Yale. He then became director of the University of Arizona's Human Energy Systems Laboratory, where he conducted extensive research with spiritual mediums. His book, published in 2002, *The Afterlife Experiments: Breakthrough Scientific Evidence of Life After Death*, detailed these experiments:

'Daring to risk his worldwide academic reputation, Dr. Gary E. Schwartz, with research partner Dr. Linda Russek, asked some of the most prominent mediums in America-to become part of a series of experiments to prove, or disprove, the existence of an afterlife. [Leading to] a breakthrough scientific achievement: contact with the beyond *under controlled laboratory conditions.* In stringently monitored experiments, leading mediums attempted to contact dead friends and relatives of 'sitters' *who were masked from view and never spoke, depriving the mediums of any cues.* The messages that came through stunned sitters and researchers alike. There are some extraordinary and uncanny revelations, and... Dr. Schwartz was forced by the overwhelmingly positive data to abandon his skepticism...'[229]

---

[229] Schwartz, G. E. PhD. & W. L. Simon. 2002. *The Afterlife Experiments: Breakthrough Scientific Evidence of Life After Death.* Atria Books.

'You know what is sad, Ali, is that so many of those who are so passionate about de-Souling and de-intelligencing Life and Existence by insisting it's all one big 'accident' and that Life after death can't, in their opinion, be possible, don't actually give the evidence that consciousness is not, as they believe, an 'accidental' Darwinian byproduct of mindless and 'accidental' atoms a chance. '

'How do you mean they don't give it a chance?'

'I mean they are simply not interested in properly studying the vast body of modern *evidence,* not religious dogma, for the reality of psychical and spiritual phenomena. In fact, when "confronted with the evidence, their reaction is often anything but rational"[230] and that they "don't want to discuss [the] evidence" as Richard Dawkins, for example, said when discussing the question of telepathy with another well known biologist, Dr. Rupert Sheldrake. Look, in Gary Schwartz's After-Life Experiments:

> 'leading mediums attempted to contact dead friends and relatives of 'sitters' *who were masked from view and never spoke, depriving the mediums of any cues.* The messages that came through stunned sitters and researchers alike.'

'We are a supposedly scientific culture. What could be more empirical and 'controlled' by way of experiment? Was this not a valid and rational way to put to the test the *scientific hypothesis* of survival of individual consciousness beyond death? Those who are skeptical about this kind of evidence *just don't want to believe the evidence.* It doesn't fit in with what they 'dream of in their philosophy' so they ignore it or demand *'yet more proof.'* They are remarkably incurious, closed-minded and unscientific in their approach and, unlike those distinguished researchers, I mentioned earlier, they refuse to look at the evidence let alone show any willingness to follow it wherever it may lead.

'Yet what could be more serious than the true nature of Life and death and whether consciousness, mind and individuality really are 'accidents' of 'accidental' bodies and therefore 'die' with them or not? As this scientist, geophysicist Ph.D. and reviewer of *Randi's Prize: What Sceptics Say About the Paranormal, Why They Are Wrong and Why It Matters* said:

> '[Psychical effects] *have* been studied in painstaking detail for 150 years by highly qualified scientists, including, in recent decades,

---

[230] Carter, C. 2010. *Science and the Near Death Experience.* Inner Traditions; p. 236.

some of the most carefully executed scientific experiments ever conducted with multi-layered experimental controls that put other fields of science to shame. Because researchers in this field are under unrelenting ... assault, they control even against absurdly improbable and unrealistic forms of cheating and fraud among other things, problems that most scientists don't have to think about at all ... Statistically and *taken as a vast body of work, their results are rock solid ...*

'The scientific facts are in; they're well-proven and extensively documented – many tens of thousands of pages of detailed studies.[231] *The demand for more proof is simply a ploy.*'[232]

'Richard Dawkins is a famous scientist and author of *The Selfish Gene, The Blind Watchmaker* and *The God Delusion*. He is a conviction materialist and campaigns against religion and belief in all phenomena that do not fit the materialist faith and belief system. In 2007, he visited Rupert Sheldrake, another well known biologist and author of *A New Science of Life, Dogs that Know When Their Owners are Coming Home, The Sense of Being Stared At* and *The Science Delusion,* to interview him for a TV series called 'Enemies of Reason' as a follow up to his 2006 attack on religion called 'The Root of All Evil'.

'Sheldrake recounts how before Enemies of Reason was filmed, the production company told him that Dawkins wished to visit him to discuss Sheldrake's research on unexplained abilities in animals and people. Sheldrake was reluctant to take part but was assured that the documentary would be "... an entirely more balanced affair than The Root of All Evil was." And that "We are very keen for it to be a discussion between two scientists, about scientific modes of enquiry". Sheldrake says:

'So I agreed and we fixed a date. I was still not sure what to expect. Was Richard Dawkins going to be dogmatic, with a mental firewall that blocked out any evidence that went against his beliefs? Or would he be open-minded, and fun to talk to? ...

'Richard began by saying that he thought we probably agreed about many things, "But what worries me about you is that you are

---

[231] See for example, Radin, D. 2009. *The Conscious Universe: The Scientific Truth of Psychic Phenomena.* Harper One.
[232] McLuhan, R. 2010. *Randi's Prize: What Sceptics Say About the Paranormal, Why They Are Wrong, and Why It Matters.* Matador. Amazon reviews by Geophysics Ph.D, Sun Dog, M. R. Barrington and Robert Perry.

prepared to believe almost anything. Science should be based on the minimum number of beliefs." I agreed that we had a lot in common, "But what worries me about you is that you come across as dogmatic, giving people a bad impression of science."

'Sheldrake then describes how Dawkins said that he'd like to believe in telepathy but maintained that there wasn't any evidence for it. He alleged that if it really occurred it would "turn the laws of physics upside down" and added that: "Extraordinary claims require extraordinary evidence." Sheldrake replied:

>"'This depends on what you regard as extraordinary. *Most people say they have experienced telepathy,* especially in connection with telephone calls. In that sense, telepathy is ordinary. The claim that most people are deluded about their own experience is extraordinary. Where is the extraordinary evidence for that?"

'Dawkins, Sheldrake says, produced no evidence in response to this question other than generic arguments as to the fallibility of human judgement and his assumption that people engage in wishful thinking when it comes to "the paranormal". Sheldrake continues:

>'We then agreed that controlled experiments were necessary. I said that this was why I had actually been doing such experiments, including tests to find out if people really could tell who was calling them on the telephone when the caller was selected at random. **The results were far above the chance level.** The previous week I had sent Richard copies of some of my papers, published in peer-reviewed journals, so that he could look at the data. Richard seemed uneasy and said, *"I don't want to discuss evidence".* "Why not?" I asked. "There isn't time. It's too complicated. And that's not what this program is about." The camera stopped.
>
>'The Director, Russell Barnes, confirmed that *he too was not interested in evidence.* The film he was making was another Dawkins polemic. I said to Russell, "If you're treating telepathy as an irrational belief, *surely evidence* about whether it exists or not *is essential for the discussion.* If telepathy occurs, it's not irrational to believe in it." ...
>
>'Richard Dawkins has long proclaimed his conviction that "The paranormal is bunk. Those who try to sell it to us are fakes and charlatans". ... But does his crusade really promote "the public understanding of science," of which he is the professor at Oxford? Should science be a vehicle of prejudice, a kind of fundamentalist

belief-system? Or should it be a method of enquiry into the unknown?'[233]

'It's disappointing, Ollie, that Richard Dawkins said to Rupert Sheldrake that he *"didn't want to discuss evidence"*. Surely scientists should always be interested in *empirical hypothesis testing* and in following the evidence… wherever it may lead.'

'You're right, Alisha, but some of the world's most influential materialist 'skeptics' are not interested in empirical evidence and anything as mundane as classical scientific hypothesis testing. This is why Rupert Sheldrake's most recent book, *The Science Delusion: Freeing the Spirit of Inquiry*, is so important. He debunks the materialist world view as being a hindrance to the progress of human knowledge. Sheldrake highlights the flaws in ten foundational beliefs that mainstream, materialistic science refuses to question. These include the idea that animals and people are complex, accidental machines, automata rather than beings in their own rights. The idea that human consciousness is an illusion. The dogma that the laws of Nature are unchanging. The non-negotiable materialist belief or superstition that Nature has no teleology, it's completely without purpose. *The Science Delusion* confirms that the 'accidental' machine metaphors beloved of materialist thinkers in biology don't make sense. As McLuhan, in his review, puts it:

> 'No machine starts from small beginnings, grows, forms new structures within itself and then reproduces itself. Yet plants and animals do this all the time and to many people – especially those like pet owners and gardeners who deal with them on a daily basis – it's *'blindingly obvious'* that they are living organisms. For scientists to see them as machines propelled only by ordinary physics and chemistry is an act of faith.'[234]

'It's not just an 'act of faith', Ollie, it's stark raving bonkers!'

'You're right, of course it is. This is what I've been trying to say. Materialism disconnects science from the Soul of all things. It is false to reality, a universally corrosive acid of untruth that strips away all meaning, intelligence and purpose from our understanding of Nature. It is a science blocking and knowledge stopping ideology.

Sheldrake explains that beyond the genes there are subtle, epigenetic

---

[233] Rupert Sheldrake's Website: Dialogues and Controversies, emphasis added
[234] Sheldrake, R. 2012. *The Science Delusion: Freeing the Spirit of Inquiry*. Coronet. Amazon review, R. McLuhan; empasis added.

formative factors at work which cause flies, mice and people to be different from each other, because their developmental genes are almost identical. These subtle formative factors may not be material at all.

'He notes that materialist thought doesn't have a clue as to what consciousness truly is and he mentions the absurd trend among some modern philosophers keen to deny the reality of their own and everyone else's minds and consciousnesses, so taken in are they by the dogmas of ideologically materialist science. Sheldrake presents evidence for telepathy and goes into the 'sense of being stared at'. This important book demonstrates how the unjustified conflation of science with the philosophy of materialism has become a brake on the progress of human knowledge. Sheldrake refutes the tottering materialist paradigm and shows that it cannot accommodate the data of reality.

'Alongside Sheldrake's *The Science Delusion,* – Chris Carter's *Science and Psychic Phenomena: The Fall of the House of Skeptics,* Robert McLuhan's *Randi's Prize: What Skeptics Say about the Paranormal, Why They are Wrong, and Why it Matters,* David Fontana's *Is There An Afterlife?: A Comprehensive Overview of the Evidence,* Michael Denton's *Evolution: A Theory in Crisis,* Michael Behe's *Darwin's Black Box: The Biochemical Challenge to Evolution* and *The Edge of Evolution: The Search for the Limits of Darwinism,* Stephen Meyer's *Signature in the Cell: DNA and the Evidence for Intelligent Design* are all important books.

These works, each in their own way, debunk the debunkers, showing them to be curiously *uninterested in the, by now, vast body of modern evidence that refutes materialism,* at times dishonest, and profoundly wrong in their unshakeable assumptions that a Cosmos of intelligence, intelligent evolution, extrasensory abilities, Life after death and multi–dimensional realities are incompatible with what we know about existence.

'Psychic phenomena, intelligent design and intelligent evolution do not violate any principles of modern physics. The extreme materialist skeptics and disbelivers in a multi-dimensional cosmos of intelligence and meaning, who have such a disproportionate influence, are *not interested in the evidence*, they refuse to seriously engage with it but are, rather, engaged in a kind of unholy war "fueled" as Carter says, "by the fervent belief that they alone are the last defenders of the citadel of science." Thankfully, most scientists are more open minded and don't identify with the extreme materialism. A 'materialism' that is, ironically, based on m—a—t—t—e—r that is not, ultimately, matter at all because it is 'made' of nothing but intelligently informed quantum-energy S—P—A—C—E – something that the spiritual–scientist, yogi–adepts of India realized centuries ago, using highly trained inner vision. As Max Planck said, "there is no matter as such," there is only intelligent force.

'Look we can all put a telescope to a blind eye and insist we can see nothing. Yet, this is so short sighted and arrogant. No one knows what the universal witnessing awareness and consciousness truly is. No one knows, in any ultimate sense, why there is anything at all rather than nothing at all. Ideological materialists cannot say why there should be endless mindlessness, unintelligence and accidents or a MUA–UGM [Mindless, Unintelligent, Accidental-Universe-Generating-Mechanism] as their supreme pseudo-uber-god and uncaused cause of all that is. All they can say is:

> *"That's just the way we believe it is.* We think mindlessness, unintelligence and accidents are supremely creative."

'Saying there was a Big Bang or that multiverses are forming every second as a way to try to make the Universe we are actually in completely meaningless is no final answer. It doesn't answer *why* there should be a Big Bang or allegedly meaningless multiverses forming every second.

'Equally, the philosophical idealists don't know *why* the Cosmos should be a multi–dimensional place of Consciousness, Intelligence and Purpose. All they can say is: *"That's just the way it is."* No more can religious and spiritually minded people say *why* God or Supreme Intelligent Beingness should exist, as the Final Source and Uncaused Cause of all that is. All they can say is: *"That's just the way it is."*

'All anyone can do is decide which of these, essentially, two versions of existence makes more sense? The unconscious, unintelligent, accidental Cosmic Mystery of materialist faith and belief or the conscious, intelligent and meaningful Cosmic Mystery of Idealist, religious and spiritual belief and experience.

'No one, however, has a right to arrogantly dismiss the large body of accumulated modern evidence that clearly there is more to existence than the kind of mindless, unintelligent and wholly accidental affair that the ideologues of Darwinian and other forms of materialism so passionately believe in.

There is no ordinary evidence anywhere, let alone the 'extraordinary evidence', of the kind that Dawkins calls for in respect to telepathy, for the bizarre and extraordinary claim that Life on Earth, the Living World, is a vast, second by second, 24/7, billions year long running *'accidental chemical reaction'!* What could be more unreasonable and non-evidence-based than that? Where is the 'extraordinary evidence' that 'accidental chemistry' is what Life is? What has our western civilization been thinking for the last 150 Darwinian years? Where is the 'extraordinary' evidence that 'chemical accidents' build or evolve anything clever at all?

'*Circular materialist arguments* that go:

'Look they 'must do', Life is here, we know it 'has to be' something random and meaningless because we believe the entire Cosmos to be one big 'accident' and there is no bearded god anywhere to be seen, not even in the most powerful telescopes, therefore *'accidental chemical reactions'* must be evolutionary!'

'won't cut it any more. People are beginning to wake up.

'The materialist skepticism that denies that consciousness can exist apart from and beyond the physical body–brain is not an open–minded skepticism but a closed–minded skepticism which insists it is 'right' and is simply not interested in 'discussing evidence'. Like Galileo's opponents, who refused to look in his telescope, it refuses to look at the evidence and dismisses it out of hand. The late Carl Sagan said 'extraordinary claims require extraordinary evidence'. So, where is the extraordinary evidence that *'accidental chemical reactions'* are 'evolutionary' and build Life Forms?

'There has never been any. It is just a materialist myth that it is 'scientific' to claim that the Living World is, literally, a Soulless, Life–Forceless *'accidental chemical reaction'* of billions of years long duration. The actual evidence is that chemical accidents cause random rust and forest fires. The actual evidence is that even if the whole universe consisted of nothing but primordial soup Life still could not arise 'by accident', it's not merely 'improbable', it's *statistically impossible.* But mainstream science and biology, that were *supposed to be about what's actually true,* not the ideology of materialism, take no notice.

14 Materialism: A Failed Hypothesis and a Science Blocker?

"Materialists often claim credit for the scientific advances of the past few centuries. But *it is the scientific method of empirical hypothesis testing*, rather than a materialistic philosophy, that has been responsible for the success of science in explaining the world. If it comes to a choice between empirical method and the materialistic worldview, the true scientist will choose the former'"[235]

We sat down on a grassy bank by the river, opening our flasks for a drink. I said,

'I have gradually come to realize, Ali, that all this is not really a matter of 'science' at all but one of heart and mind. A science captured by the ideology of materialism cannot accept the evidence for intelligence and intelligent evolution in the Living World or for Life after death and for psychic phenomena such as telepathy not because the evidence is not valid but because it *refutes the materialist paradigm*. Such a science has given up on classical hypothesis testing and empirical truth. It has become dogmatic. It has ceased to be genuinely scientific.

'If one's mind is closed then it doesn't matter what the evidence is. Someone could show us a piece of moon rock, even hit us over the head with it, but we could continue to insist that it's a dairy product. We can all believe what we want to believe quite regardless of the evidence.

'Some people think the purpose of 'science' is to use the immaterial Soul faculties and spiritual essences that we call *mind* and *intelligence* to explain how *mindless* and *unintelligent* Nature and Existence are, from atoms to galaxies; how the Cosmos is meaningless, it's an 'accident'. Gravity, Light, Electricity? All accidents. The stunningly smart-materials-chemicals? Accident. DNA CODING "more advanced than any software"? Accident. Therefore, the amazing and magical Soul Life Form Beings of the Plants, Animals and People must all be 'accidents' as well and there can be no true Mind, Soul or Life after death.

---

[235] Greyson, B.: "Commentary on 'Psychophysiological and Cultural correlates Undermining a Survivalist Interpretation of Near-Death Experiences.'" *Journal of Near Death Studies* 26, no. 2 (Winter 2007). p. 142.

'Yet the evidence that Life after death is a fact of existence is in and has long been in.[236] If mainstream, materialist thinking science refuses to consider the evidence, let alone to accept it, *it is because it doesn't want to* – it is not for 'lack of evidence' but for psychological reasons. For reasons that, ultimately, baffle me, Ali, the view seems to be that it is just somehow more 'scientific' to imagine that Life is a meaningless, unintelligent 24/7, second by second *'accidental chemical reaction'* that has continued for billions of years for no intelligent rhyme or reason at all. It is just more 'scientific' to assume, *with not a thimbleful of evidence mind you,* that the trillions of stunningly complex biochemical operations taking place in everyone's bodies right now, this very second, every second, 24/7, 7/52 are the results of utterly unintelligent, purposeless and accidental processes. If that is not a non-evidence-based delusion or collusion I don't know what is.

'For modern materialist thinking subscribing to *an unintelligent and meaningless Mystery of Existence,* strangely, seems to appear more meaningful and intelligent than subscribing to *an intelligent and meaningful Mystery of Existence.*

'The materialist view, Ali, makes our science irrational. We are caught between a rock and a hard place. Religions who insist that their concept of existence based on ancient and unyielding dogmas and promises of strange heavens and threats of hellish hells is the only correct view and a once Idealist, 'mind-before-mechanism' western science kidnapped by de-purposing materialist thought which regards existence as 'mindless and unintelligent' (which is self-refuting) and 'accidental' (which is absurd).

'One thing, Ollie . . .'

'Yes . . .

'Why do you care so much about this. Why does it matter to you so much whether or not it's all one big 'accident', whether we survive the death of the body and so on?'

'It's partly because I like schemes of scientific and philosophical understanding that are coherent and match the data of reality. I also like schemes of thinking that are based on evidence. *Do 'accidental chemical reactions' build or evolve 'machinery' of any kind, simple or complex?* Where's the extraordinary evidence for that extraordinary claim? Where is the evidence that Consciousness, Sentiency, Mind,

---

[236] Fontana, D. 2005. *Is There an Afterlife?: A Comprehensive Overview of the Evidence.* O Books. Review, product description, Amazon. Schwartz, G. E. PhD. & W. L. Simon. 2002. *The Afterlife Experiments: Breakthrough Scientific Evidence of Life After Death.* Atria Books.

Soul and Intelligence are all the 'accidental' byproducts of atoms of matter that is itself, in materialist thought, considered to be utterly mindless and unintelligent?

'That does not make sense, it is incoherent and self-refuting – either (a) matter itself is inherently intelligent in various ways, that is matter is itself pervaded by Mind, Intelligence and Consciousness and therefore can express them, a view that is sometimes called panpsychism, or (b) matter is 'dumb' in which case 'mindless' matter + Consciousness and Intelligence combine together to give rise to the Soul–Life–Form Beings of the amazing Plants, Animals and People. Neither of these positions are 'materialist' view points. Both are idealist because they accept consciousness, intelligence, motivation and purpose as real and as ontological properties of the Cosmos. Materialism, ironically, uses Mind and Consciousness to assert that Consciousness is not truly real, it's an 'accident of matter' and, in view of this dreary (and self-refuting) conclusion, holds that:

"We're all zombies. Nobody is conscious."[237]

'Apart from the self-refuting nature of this view the classical intuition and rational inference that there is more to existence than 'accidents', all the way up and down, is obviously more consistent with the evidence of reality, when taken in the round.

'I also like the classical intuition and spiritual realization that numbers, mathematics and geometry mean more than materialist thought allows us to ascribe to them. Pythagoras was onto something when he talked about the inner meanings of numbers. The dimensions of existence really are expressions of sacred geometry. I like the ancient knowledge, that some indigenous people still have, that the Living World is teeming with other spiritual beings, Nature Spirits, Devas, Elementals, Sylphs and Undines, Fairies and Angels even if only a few of us, with sufficiently developed sensitivities and sensibilities, can see or experience them.[238] I find the modern evidence that Angels are real and are, at times, at least, able to help those, in need, from subtle levels and in subtle ways, interesting and inspiring.[239] I like the Idea that we Live in an intelligent, multidi-

---

[237] Dennett D. 1992. *Consciousness Explained*. Back Bay Books, p. 406. Quoted by L. Dossey: 2010. Is *the Universe Merely A Statistical Accident? The Blog. Huffington Post.*
[238] Tompkins, P. 1997. *The Secret Life of Nature: Living in Harmony with the Hidden World of Nature Spirits from Fairies to Quarks*. Thorsons.
[239] Byrne, L. 2010. *Angels in my Hair*. Arrow. McMahon, P. 2011. *Guided by Angels*. Collins.

mensional, multi-vibrational Cosmos of Meaning, perhaps even one of Love – even if that Love is, on the surface of things, often so very hidden: a heart of Love that beats behind all the outer seeming which can be so unbelievably harsh.

'We think it's 'okay', Ali, to fight wars, bomb others to death, do terrible experiments of Living Animals. I am baffled by a Cosmos in which such things are possible, or one that gives rise to rasping parasites and any number of ghastly diseases, but that does not make the Universe meaningless or unintelligent. It makes it mysterious, challenging and baffling, perhaps involving battles between different kinds of inter–dimensional intelligences some benign and other malign, who knows why? Why do some people enjoy causing immense stress and distress around the world by designing computer viruses?

'Personally, I believe that, ultimately, despite all the appalling things that we sometimes do, we are children of meaning, love and intelligence, not 'mindlessness, unintelligence and accident', no matter how weird and challenging our existences are and no matter how challenging we make them.

'Right now, though, we have a materialistically oriented science that is not interested in the Soul of things, the interiority of existence, the blazing, intelligent Light at the heart of reality. Rather it is a science that is obsessed with matter and mechanism to the exclusion of all mind, Soul, feeling, intelligence and meaning. A science that understands the outer mechanisms of many things in Nature but the intelligence, true meaning, Life and Soul of nothing in Nature – because it insists that there are none! It even attacks and mocks those few scientists brave enough to risk its ridicule by making the modest (yet blazingly obvious) claim that many features of Nature, especially Living Nature, look intelligent and intelligently designed or intelligently evolved because they really are! Courageous thinkers like Denton, Davidson, Behe, Dembski, Meyer, Latham, Gonzalez, Wells, Axe and many others.

'The 'scientific' materialist view which makes it into a career damaging crime to challenge its cherished but absurd pseudo-uber pantheon of 'mindlessness', 'unintelligence', 'accident' and 'unmeaning' as its supreme, supremely unreasonable and, in fact, impossible causative agencies in Nature is not only self-refuting, given that we emerge out of Nature, and we are full of intelligence, love and purpose, as well as, at times, darker tendencies, but also Soul destroying – because it seeks to suck the meaning, the heart and Soul out of everything with its irrational (pseudo) 'rationalism', (where 'rational' is falsely defined as meaning the materialist world-view only) and its, at times, heartless and de-Souling form of knowledge gathering.

'How *can* 'science' have any meaning if the Cosmos is unintelligent and meaningless, Alisha? Where is the extraordinary evidence for that surreal claim? Do you see how absurd that is? Can you not see how self-refuting is the notion that we can make intelligible and find meaning – that is find out how things intelligently work, how they are cleverly put together – in an 'unintelligent' cosmos? $E=mc^2$ is 'unintelligent' in the materialist scheme. Einstein is an 'unconscious zombie' in the materialist scheme. The stunningly precise fine tuning of the laws of Nature are 'purposeless' and 'meaningless' in its de-Souling and de-intelligencing scheme. The astonishingly clever, 24/7 functioning of Living Cells is unintelligent and lacking in teleology in the counterfactual materialist scheme.

'So where does the faculty, we call *intelligence* – that permits us to discover how, allegedly, utterly 'unintelligent' is the total Natural matrix that we come out of – itself come from? From outside the known Universe perhaps? Does COBOL emerge from *'mindlessness' and 'unintelligence'?* But DNA CODING, vastly in advance of COBOL supposedly does? This is why I find materialist thinking so unreasonable, Ali. It is intellectually incoherent and either deluded or dishonest. Yet, strangely, it has fooled people, for more than 150 long years now, into thinking it's the rational and respectable view.

'How could we object to the materialist, Darwinian, unintelligent design world views if the way we made Watches and Cars and Computers was by putting fragments of rocks and earth in barrels and tumbling them, perhaps labeling one barrel 'Watches', one barrel 'Cars' and one 'Computers', setting them on a tumble, then, a while later, opening them and taking out the finished products. If that was the way we made cars and watches and computers we'd have to agree that that is probably the way Nature makes digital DNA CODES, 'far, far more advanced than any software ever created,' Living Cells and Life Forms too – mechanisms that are far more complex than any car, watch or computer.

'Saying that's only an analogy, and therefore, according to Hume, is false, is degrading to reason. If we cannot use arguments from analogy we can no longer reason at all. To say we are not comparing precisely like with like is correct. Living Cells and the larger Life Forms they help to build are many orders of magnitude more complex than any 'machinery' we make. This is what is also so depressing about the materialist world view, Ali, it lacks intellectual integrity.

'One minute it is insisting that all the mechanisms of existence from atoms to molecules to plants to animals to people to planets to stars are just Soulless 'mechanisms', 'accidental machinery'. However, if you point out that 'machinery', simple or complex, never arises 'by accident' then your 'design argument' is suddenly invalid because you are arguing from analogy. Yet, we have no real–world, let alone 'extraordinary' evidence that 'accidental chemistry' is what Life and the Life Form Soul Beings actually are.

'This seems like a beautiful and meaningful existence to me, Ali. Yes, it has many deeply challenging and baffling aspects, a good many of which we ourselves create, but to insist that it's all just one big 'accident' because it is so hard or because we don't believe in some absurd god-with-a-beard making and sustaining it all seems sad.

'We worship a 'science' that hasn't a clue what sacred, magical, Living, second by second by second Life is, a science that denies the existence of a Life Force or Life Forces and the Soul and Spirit in all things partly because someone artificially synthesized urea in 1828! A science that refuses to take any notice of the accumulated and continuously mounting evidence that Consciousness is not an 'accident' of an 'accidental' body and 'accidental' brain (an accidental 'computer made of meat' with more molecular-scale switches than *all* the computers and internet connections on the entire planet), that Life and Mind and Soul do survive the chemical 'death' of the body.

'We admire a materialist science that uses its own Soul imbued powers of *consciousness, mind and intelligence* to insist that Life is a Soulless, rhymeless, meaningless, zombie, second by second 24/7 *'accidental-chemical-reaction'*. A science that equates 'rationality' with the surreal and scientifically impossible idea that the most complex 'bio-machines' we know of, that make our supercomputers look like babies rattles, arose by 'chance' and just as surreally, perhaps even more surreally, that they are working 'by accident' right now, blazingly clever microsecond by blazingly clever microsecond – trillions of astonishingly complex bio-chemical operations taking place in our bodies this very instant without the involvement of any superordinate Life–Forces or Soul–Forces anywhere to be allowed or inferred. What for? No reason at all, it's all just an 'accident'. In the de-Souled and de-intelligenced materialist Cosmos of utter unmeaning and pointlessness, it's mindlessness, unintelligence and 'accidents' or MUAs all the way up and all the way down, with no ifs or buts or any reasonable get outs or intelligent exceptions of any kind at any time.

'A shrunken, mono-dimensional Cosmos in which nothing anyone ever says *actually means* anything at all or *can* mean anything at all, including the pursuit and the pronouncements of science, because it's *all* [allegedly] mindless and *unintelligent*. In the peculiar materialist scheme the concepts, let alone the ontological *essences,* of *intelligence* and *purpose* cannot exist, because if Nature has *no intelligence* then neither do we and our science, like everything else, becomes meaningless. A Cosmos in which:

"We're all zombies. Nobody is conscious."[240]

'This is where materialism takes us. And, supposedly, it's a rational world view. You know, Ali, as Fred Hoyle said:

'... the probability of life originating at random is ... absurd ... It is therefore almost inevitable that our own measure of intelligence must reflect ... *higher intelligences* ... even to the limit of God ... such a theory is so obvious that one wonders why it is not widely accepted as being *self-evident*. The reasons are *psychological* rather than scientific.'[241]

'We use our *minds and intelligence* to fight and deny the *intelligence and purposefulness* of our own minds and the *intelligence and purposefulness* that pervades the Living World, the 24/7 *intelligence and purposefulness* that drives the tiny Living Cells, and the magical and Soulful Life Forms of the Plants, Animals and People and the seasons and the winds and the oceans and the turning of the planets and the stars not because our 'science' tells us to but, as Hoyle says, for psychological and emotional reasons. It's a shame, but that's how it is.

'You know, Alisha, I sometimes doubt that *any amount* of 'evidence' will make any difference. F. W. H. Myers' cross-correspondences and Gary Schwartz's After–Life experiments, for example, should be all the evidence we need, quite apart from all the other countless experiments and demonstrations, over the last 150 years, showing the same thing, that far from being an 'accident' or 'unintelligent' and 'purposeless' or lacking in teleology, total Holistic Nature is blazingly, dazzlingly intelligent and contains 'super', higher Natural or multi–dimensional components, where intelligent Consciousness continues beyond or parallel to this World, including Consciousnesses who were once here, but if those who 'speak for science' today refuse to engage with the extensive research, and say that they *"don't want to discuss evidence"*, insisting instead that, "No, you're wrong Mr Galileo, the Sun orbits the Earth and we're the majority" then who are we to argue with them?

'If some courageous scientists say, as they repeatedly have, that the *scientific* (not 'religious') evidence that we Live in a Cosmos of

---

[240] Dennett D. 1992. *Consciousness Explained*. Back Bay Books, p. 406.
[241] Hoyle, F. and N. C. 1981. Wickramasinghe, *Evolution from Space*. J. M. Dent & Sons.

Meaning and Intelligence is there, it's *indisputable and irrefutable,* that our amazing Living World is no 'accident', challenging and weird and mysterious, yes, but no 'accident', but their colleagues refuse to take any notice of them for whatever 'psychological' reasons then what are we to do? Like poor Galileo "e pur si muove" Galilei all we can do is mutter under our breaths, "Viviamo in un Cosmo di intelligenza e significato, la vita non è un accidente; e la Vita dopo la morte è un fatto accertato oltre ogni ragionevole dubbio."

'As David Fontana says:

"Ultimately our acceptance of the reality of survival [beyond 'death'] may not come solely from the evidence but from personal experience and from some inner intuitive certainty about *our real nature. We are who we are,* and at some deep level within ourselves we may be the answer to our own questions. If your answer is that *you are more than a biological accident* whose ultimately meaningless life is bounded by the cradle and the grave, then I have to say I agree with you."[242]

'Dr Peter Fenwick, neuro–psychiatrist and the UK's foremost clinical authority on Near-Death-Experience says of David Fontana's work:

'Scholarship, personal experience and high quality always show. This highly accessible, detailed and authoritative book will become a classic. After reading it and assessing the evidence, **there can no longer be any doubt that there is life after death**. David Fontana's book should be mandatory reading for all those involved in the care of the dying – and of course for the rest of us who know we will have to face it one day!

'The spiritual traditions of both West and East have taught that death is not the end, but modern [materialist] science generally dismisses such teachings. The fruit of a lifetime's research and experience by a world expert in the field, *Is There An Afterlife?* presents *the most complete survey to date of the evidence,* both historical and contemporary, for survival of physical death. It includes mediumship and channelling, spontaneous cases, hauntings, apparitions, near death experiences, out of the body experiences, EVP or Electronic Voice Phenomena, Instrumental Transcommunication and recent laboratory work.

---

[242] David Fontana, Ph.D. A professor of transpersonal psychology in the UK. Dr. Fontana is a past president of the Society for Psychical Research and a fellow of the British Psychological Society. He has done extensive survival research and is the author of many books, including *Is There an Afterlife?: A Comprehensive Overview of the Evidence.*

'It looks at the question of what survives – personality, memory, emotions and body-image in particular – exploring the question of *consciousness as primary to and not dependent on matter* in the light of recent brain research and quantum physics. It discusses the possible nature of the afterlife, the common threads in Western and Eastern traditions, the common features of "many levels", group souls and reincarnation. ... giving due weight to the claims both of science and religion, *Is There An Afterlife?* brings it into personal perspective. It asks how we should live in this life as if death is not the end, and suggests how we should change our behavior accordingly.'[243]

'Let's give the last word, Ali, on high quality psychic ability, mediumship and inter–dimensional, telepathic communications providing evidence of Life after Death and proof that Consciousness and Mind are not random Darwinian 'accidents' of matter to a great and brilliant mind, that of Sir Arthur Conan Doyle, 1859–1930. Doctor turned writer, creator of Sherlock Holmes. Initially, highly skeptical of psychic phenomena, he said:

> "Healthy skepticism is the basis of all accurate observation, but there comes a time when incredulity means either culpable ignorance or else imbecility, and this time has been long past in the matter of spirit intercourse."

'Sir Arthur Conan Doyle wrote those words a long time ago now. If our materialistically oriented scientific culture still prefers to disbelieve the accumulated modern evidence for the fact that Consciousness, Sentiency and Mind are neither produced by matter nor are they 'accidental' epiphenomena of it but are merely expressed and transmitted by it, by stunningly clever processes, then it is not for reasons of evidence but of psychology. Although physical Life comes to an end for all of us our supposedly scientific culture seems remarkably incurious as to what lies beyond.

'There is, now, a great deal of evidence as to what lies beyond 'death', which confirms, at the minimum and beyond all reasonable doubt, the survival of individual consciousness. One of the terribly damaging consequences, however, of materialist thought is that for it to remain apparently true it must ceaselessly deny, as it ceaselessly does, vast amounts of Lived human experience that points to a multi-

---

[243] Fontana, D. 2005. *Is There an Afterlife?: A Comprehensive Overview of the Evidence.* O Books. Review, product description, Amazon.

dimensional Cosmos of intelligence, meaning and purpose rather than a mono-dimensional one of mindlessness, unintelligence and accident. As noted before, the psychic dimensions of Life:

'have been studied in painstaking detail for 150 years by highly qualified scientists, including, in recent decades, *some of the most carefully executed scientific experiments ever conducted* with multi-layered experimental controls that put other fields of science to shame. . . . *their results are rock solid* . . .

The demand for more proof is simply a ploy. [244]

'Most of those thinkers, today, who are so quick to rubbish and mock the sea of evidence and experience, including a great deal of serious scientific research, that there really is more to Reality than meets the closed-minded, materialist-skeptic eye, *haven't studied these matters* and refuse to do so[245] by contrast with those bright and open minds like Sir William Crookes, Sir Oliver Lodge and Sir Arthur Conan Doyle who did do so over a century ago and Dr. Gary Schwartz, David Fontana Phd., Dr. P. Fenwick, Dr. Rupert Sheldrake and many others more recently.

'It is easy to debunk and mock that which doesn't interest us or in which we do not believe. True science, Ali, both physical–science and spiritual–science, of which the world's religions and spiritual traditions are all parts, can only progress in a spirit of genuinely open–minded inquiry and in empirical hypothesis testing not in one of closed–minded preconception and dogma – be they the materialist dogmas of a 'pointless and random' Cosmos of 'unintelligence and unmeaning' or the unexamined religious dogmas of, for example, just one arbitrary human Life, (without, apparently, any rhyme, reason or justice), then an eternity of a bizarre 'heaven' or 'hell'.

'Yet total Holistic Reality is so much more amazing than either materialist science or orthodox religion allows it to be. As this near death experiencer affirms:

'being bathed in The Light on the other side of death is *more than life changing*. The Light is the very essence, the heart and soul, the

---

[244] Geophysics Ph.D, Sun Dog, Amazon reviewer of *Randi's Prize: What Skeptics Say About the Paranormal, Why They are Wrong, and Why it Matters*.
[245] Radin, D. 2009. *The Conscious Universe: The Truth of Psychic Phenomena*. Harper One. Carter, C. 2012. *Science and Psychic Phenomena: The Fall of the House of Skeptics*. Inner Traditions. McLuhan, R. 2010. *Randi's Prize: What Sceptics Say About the Paranormal, Why They Are Wrong, and Why It Matters*. Matador.

all-consuming consummation of ecstatic ecstasy. It is a million suns of compressed love dissolving everything unto itself, annihilating thought and cell, vaporizing humanness and history, into the one great brilliance of all that is and all that ever was and all that ever will be. . . .

'You can no longer believe in God, for belief implies doubt. There is no more doubt. . . . And you're never the same again.

'And you know who you are . . . a child of God, a cell in The Greater Body, an extension of the One Force, an expression from the One Mind. No more can you forget your identity, or deny or ignore or pretend it away. . . . The Light does this to you. . . . And you melt away as the 'you' you think you are, reforming as the 'YOU' you really are, and you are reborn because at last you 'remember'.

'Although not everyone speaks of God when they return from death's door as I have here, the majority do. And almost to a person they begin to make references to oneness, allness, isness, the directive presence behind and within and beyond all things.

'Down through the ages this kind of knowledge has been termed *enlightenment* – literally a waking up to light, . . . a reunification with The Light. And there are groups, [religions], isms and schisms, that decree how one can reach such a state of enlightened knowingness. The rules are many, the pathways numerous, yet the goal is always the same . . . reunion with the source of your being, God.'[246]

We walked on quietly for a while, before I spoke again.

'Modern quantum physics has shown that the Universe is not, ultimately, a truly 'material' place at all, it is rather, as the yogi–adept, spiritual–scientists of India always held, a vast and extending expression of intelligent Beingness and Cosmic Consciousness,[247] the whole thing alive, intelligent, brilliant and scintillating, containing many dimensions and vibrations of existence, gross and subtle.

'This all means that the Darwinian explanations of Life as 'accidental chemistry' or materialist biology's wish to reduce the Living World to an accidental epiphenomenon of physics[248] do not make sense and are unnecessary. Basically, our choices are simple:

---

[246] Atwater, P. M. H. 1994. *Beyond the Light: Near Death Experiences.* Thorsons, p. 152.
[247] Goswami, A. 1993. *The Self-Aware Universe: How Consciousness Creates the Material World.* Jeremy P. Tarcher. Capra, F. 1992. *The Tao of Physics.* Flamingo.
[248] "The ultimate aim of the modern movement in biology is to explain all biology in terms of physics and chemistry" Crick F. 2004. *Of Molecules and Men.* Prometheus, p. 10. Quoted by L. Dossey: 2010. *Is the Universe Merely A Statistical Accident? The Blog. Huffington Post.*

(a) Mindless and Unintelligent Cosmic Mystery that 'accidentally' gave and gives rise, right now, second by second, supposedly for no reason at all, yet, continuously, by means of trillions of stunningly complex operations, 24/7, to our Bodies, Minds and Intelligences or

(b) Intelligent and Meaningful Cosmic Mystery that gives rise to our Bodies and permits the transmission of our Minds and Intelligences into this physical dimension for Soul experience, learning and growth, blazingly intelligent second by scintillatingly dazzling second.

'We cannot escape *the Great Mystery of Existence,* for no one knows, not even the great yogi-adepts and the spiritual masters, *why* there is anything at all rather than nothing at, all we can do is decide what kind of multi–dimensional Mystery we think we are in – an intelligent one or a meaningless one. The choice we make is not one of science as such, for, if it was, the choice would be obvious, but also one of heart and mind, mood and feeling. If we are to use pure logic and reason then, res judicata, the matter was decided long ago by some of the greatest minds that have ever lived, Pythagoras, Socrates, Plato, Aristotle and Aquinas to name but a few – intelligent mechanisms only ever have intelligent Sources.

'These questions, Ali, are not, ultimately, a matter of 'science' but of psychology and mood, of what we want to believe. If we think or feel it makes more sense to use, in our pursuit of science, the immaterial, Living quality that we call *intelligence* – (after all, what is *intelligence,* if we think about it, as an *essence,* an ontological *quality,* in its very *thinginess?* What is its real *nature?* What does it *consist in?* How do we seize and understand *intelligence,* except by means of it's own nature, by means of its own self-existing, brilliant, scintillating *essence?*) – to continue to insist that 'mindlessness leads to accidental minds', 'unintelligence leads to accidental *intelligence',* 'accidents build the most sophisticated (bio) 'machinery' we know of' without any extraordinary evidence at all for these extraordinarily unreasonable and irrational claims apart from (a) the never proven claim that 'Darwin showed Life on Earth to be an *'accidental chemical reaction''* and (b) the belief that the immateriality and consequent physical invisibility of Life's 'higher' Natural causes means that there simply are none, then we can.

'We know, however, that, although they too are immaterial, *ideas are real.* We don't need a physical triangle or cube to prove the existence of the ideas, triangle and cube. We know that a Car is first an immaterial concept before it is a 'form-alized', 'material-ized', 'mechan-ized' object – with a purpose – one that gives visible form to all four of Aristotle's famous causes that all things have. Materialist thinking, however, denies the universal and inescapable applicability of Aristotle's four causes to phenomena in Nature and, therefore, refuses to bestow this same capacity to give expression to ideas, intelligence and purpose on the amazing multi-dimensional Cosmos from which we with our *immaterial ideas, concepts, inventions* and *immaterial intelligences* so astonishingly and *intelligently* emerge, insisting, instead, that it is a Cosmic Mystery of 'mindlessness, unintelligence and accident'.

'Come on, though, Ollie, let's be a bit more upbeat! Isn't it interesting and exciting that some really outstanding thinkers have thoroughly investigated the whole issue of human minds and consciousness as something apart from the brain and concluded that the evidence that we live in a multi–dimensional, multi-frequency universe of *Intelligence* and *Consciousness*, a mental place, is real, just as the world's earlier peoples, today's indigenous peoples and the yogis and sufis, the saints and sages have always held.'

'Yes, you're right. It is exciting and interesting. There really is a great deal of information available today, which, when taken all together, amounts to *overwhelming evidence* that bio-chemical death (the break down of the TV or the Body) is not the end of the 'show' (the individuality, mind and Soul) and that the materialist belief that total Nature is just a big 'accident' makes no sense. The modern idealist, multi-dimensional and spiritual–scientific world views are based not on one but on many pieces of evidence, amassed in a body of serious and extremely careful research, over 150 years worth now, that all point to the primacy of consciousness, mind and intelligence over 'matter' which, after all, as the yogis always said is itself, ultimately, mind-stuff too, or chitta.

'It is the *accumulated* evidence of modern research that confirms the nearly universal human realization that consciousness is more than the body,[249] let alone an 'accident' of the body, any more than TV Shows are 'accidents' of TVs that are themselves 'accidents'.

'Important parts of the cumulative evidence for the reality of consciousness and mind apart from the chemical body are the many accounts of near death and out of body experiences now formally documented and studied by bodies like the International Association for Near Death Studies or IANDS. Dr Raymond Moody's *Life After Life*,

---

[249] Kelly, E. F. & E. W. Kelly, A. Crabtree, A. Gauld 2009 *Irreducible Mind: Toward a Psychology for the 21st Century*. Rowman and Littlefield.

Margot Grey's *Return Form Death,* Dr Peter Fenwick and Elizabeth Fenwick's *The Truth in the Light: An Investigation of over 300 Near Death Experiences,* Dr Fontana's *Is There an Afterlife?,* Dr Pim van Lommel's *Consciousness Beyond Life* and Chris Carter's *Science and the Near Death Experience* and his later book *Science and The Afterlife Experience* are just a very few of the books that document these phenomena.

'Evidence that makes the idea that we 'have' to explain Life's evolution on this planet as a 'freak thing', because of the materialist dogma that we Live in a mono–dimensional Cosmos of 'mindlessness, unmeaning and accident' just does not make sense. Evidence that tells us that mind and consciousness are not, after all, the 'accidental' byproducts of 'accidental' matter as materialism holds.

'Dr Pim van Lommel, a well known cardiologist, was so struck by the accounts of their near death experiences that some patients shared that he decided to risk his reputation with a systematic, clinical study of the near death phenomenon. Over a twenty year period he interviewed hundreds of heart patients who had all *clinically died,* some for five minutes or longer, before being resuscitated. Of these a significant percentage reported an ongoing experience after the medical monitors had reported them to be clinically dead. Half were aware that their bodies were dead and some recalled witnessing the actions of hospital staff around their corpses from out of body perspectives.[250] Dr van Lommel published the results of his work in the Lancet, in 2001. He says that the results:

> 'demonstrated that common [materialist] hypotheses that dismiss the NDE as a real event, such as a lack of oxygen to the brain and fear of death had no correlation to the occurrence of an NDE.'[251]

'Darwinian–materialism holds that consciousness and intelligence are the 'accidental' creations of an 'accidental' brain. In *Science and the Near Death Experience* Chris Carter refutes this view. Examining ancient and modern accounts of NDEs from around the world, including China, India and tribal societies such as the Native American and the Maori, he explains how NDEs provide evidence of consciousness surviving the death of the body. Robert Bobrow MD, Clinical Associate Professor of Family Medicine at Stony Brook University, says of Carter's book:

---

[250] Lommel, P. 2010. *Consciousness Beyond Life, The Science of the Near Death Experience.* Harper One, back cover; emphasis added.

[251] Gustus, S. 2011. *Less Incomplete: A Guide to Experiencing the Human Condition Beyond the Physical Body.* O Books; from foreword by Dr. Pim van Lommel.

'The belief that consciousness itself is somehow produced within the brain will topple under the momentum of observations this [materialistic] theory simply cannot explain. ... A readable, informative, and *devastating critique of materialism.*

'Mario Beauregard, PhD, Professor of Neuroscience, University of Montreal, and coauthor of *The Spiritual Brain,* says of Carter's book:

'... *the evidence* does not support the mainstream scientific view that consciousness and mind are produced by the brain. ... [It] objectively reviews the empirical data on near-death experiences and rightly concludes that these data fully support the notion that mind and consciousness can continue to operate after the cessation of brain activity.

'Ervin Laszlo, author of *Science and the Akashic Field* and founder of the Club of Budapest, said:

'There has been a spate of books on the afterlife and the immortality of consciousness lately, indicating a resurgence of interest in what is surely one of the most important – and I would argue THE most important – question a conscious human being can pose in his or her life. Carter's book is not only an important contribution to this literature; it is its current crowning achievement. For he masters both the theoretical and the evidential approach, showing that *belief to the contrary* of the survival of consciousness is mere, and now entirely obsolete, *dogma,* and that the *evidence for survival is clear and rationally convincing.* A book to read and remember for the rest of one's life.'[252]

'In *Science and the Near-Death Experience* Carter assesses the modern evidence that human consciousness and mind are not dependent, ultimately, on the brain and can exist quite apart from it. After considering evidence from neuroscience, quantum physics, theories of memory and the nature of Life, the very principle that animates, informs and organizes the Life Forms, Carter concludes that the empirical evidence and the known laws of science can fully accommodate the transmission or filter theory of mind. He goes on to explore and refute, one by one the various materialist proposals for

---

[252] Carter, C. 2010. *Science and the Near Death Experience.* Inner Traditions; reviews by Bobrow, Beauregard, Laszlow, back cover.

explaining away NDEs as not being what they appear to be. As this reviewer of the book says:

> 'These chapters were a real tour de force. He examined each of a dozen proposed explanations in detail, finding in each case that the phenomenon that supposedly explains NDEs (e.g., dissociated states, oxygen starvation, ketamine) is simply not a good match for the actual characteristics of NDEs. By the time he is done, all of the proposed [materialist] alternative explanations look so weak and flimsy that *they appear to really rest on the underlying confidence that a materialist explanation simply must be true.* '[253]

'This is why materialism in science, Ali, is a science blocker and stopper. It causes mainstream science to ignore all evidence that shows materialism to be false. The trouble is that, as Carter says:

> 'In our modern world, science and scientists hold a great deal of prestige, and so few people want to be thought of as unscientific. To be labeled unscientific is enough to have one's work dismissed from serious consideration by the academic establishment. If to be scientific is good and unscientific bad, and if the term "scientific" is thought to be synonymous with the term "materialistic," then any talk of disembodied minds or spirits is anti-materialist, unscientific, and therefore bad.[254] **The long-standing confusion of materialism with science** is what largely accounts for the persistent social taboo responsible for the ignorance and dismissal of the *substantial amount of evidence that proves materialism false.* '[255]

'Bruce Greyson, a world renowned expert in the field of near-death studies, says of the science damaging conflation of science with the ideology of materialism that:

> "Materialists often claim credit for the scientific advances of the past few centuries. But *it is the scientific method of empirical hypothesis testing,* rather than a materialistic philosophy, that has been responsible for the success of science in explaining the

---

[253] Carter, C. 2010. *Science and the Near Death Experience.* Inner Traditions; Amazon review by Robert Perry; emphasis added.
[254] Carter: "For the materialist, the term unscientific seems to be the modern equivalent of the term heretical, and it is invoked for the same purpose: to exclude from consideration ideas that challenge the believer's faith."
[255] Carter, C. 2010. *Science and the Near Death Experience.* Inner Traditions; p. 238.

world. If it comes to a choice between empirical method and the materialistic worldview, the true scientist will choose the former'"[256]

'Ideological materialism, however, causes many scientists to abandon empiricism and to opt instead simply to ignore all evidence that falsifies its paradigm. This is why it is a science blocker and stopper and unscientific, even anti-scientific. Robert Perry concludes his review:

> 'In the end, the *common equation of science with materialism comes out looking like an ideology,* like its own kind of dogmatic faith.
> 'This deserves to become a landmark book in the survival debate. Carter has a real gift for presenting complex, technical issues in simple, layman's terms. And he has an even more impressive gift of total fearlessness in the face of prevailing dogma. He never flinches, yet he meets this dogma, which depends so heavily on ridicule, without ridicule of his own. His arguments have the feel of a Zen swordsman, dispassionate but deadly accurate.
> 'I am simply glad that Carter is out there writing. His book shows that those who believe in survival do not have to apologize, be timid, or take refuge in the mystery of "faith." *On strictly scientific grounds, they are in the stronger position.* With more books like this one, our society may start slowly waking up to this fact, with all its immense implications.'[257]

'Another reviewer said of Carter's book:

> 'William James has a well-known aphorism: It takes but a single white crow to demonstrate the non-universality of the contention that "all crows are black." ...
> 'Now, this hypothesis should in principle be easy to falsify. It should, according to James, take but a single case of "mind without brain (-activity)" to demonstrate the non-universality of the [materialist] contention that [as Francis Crick put it] "you're nothing but a pack of neurons." In principle, this is so. In practice, however, it will take a whole flock of white crows to make such a provoking

---

[256] Greyson, B.: "Commentary on 'Psychophysiological and Cultural correlates Undermining a Survivalist Interpretation of Near-Death Experiences.'" *Journal of Near Death Studies* 26, no. 2 (Winter 2007), p. 142.

[257] Carter, C. 2010. *Science and the Near Death Experience.* Inner Traditions; Amazon review by Robert Perry; emphasis added.

falsification. I don't know how big the flock must be, but there is surely *a critical mass* where the cases of white crows have become so numerous that *they can no longer be ignored*. The flock of NDE/OBE-cases, many of them including *verifiable out-of-body observations,* seems to be approaching that level with an accelerating pace.

'... *Science and the Near-Death Experience* ... is certainly, as Bruce Greyson has put it, "the best book on NDEs in years." Its strength is (among others) that it takes all 'skeptical' explanations seriously, examines them thoroughly and *demonstrates why they all fail.* The necessary [and most parsimonious] conclusion is that NDE/OBEs are, in fact, what they always seemed to be, and what all experiencers hold them to be: Experiences of a mind, which has left its body.

'Perhaps you will insist that "science" cannot accept this conclusion, because it must adhere to a *materialistic monism?* Well, what is the task of science? Is it to explain phenomena or to explain them away, following a pre-set ontology? Is the ontological basis for science testable or not? If it is not, that basis is in effect a dogma. And science is no longer science.'[258]

'In *Recovering the Soul: A Scientific and Spiritual Search,* Dr Larry Dossy, MD discusses the following case. Sarah was a young woman who during a routine gallbladder operation experienced cardiac arrest and died. After her body was resuscitated she was able to describe events around her body, during the period that she was clinically dead, in vivid detail. One of the corroborating pieces for the truth of her experiences and that she was really existing as a consciousness outside of her now dead body was that she noticed that the anesthesiologist was wearing mismatched socks. The skeptical retort is that she must have caught a glimpse of the socks on her way into theatre. Sarah has, however, been blind since birth.[259]

In *Mindsight: Near-Death and Out-of-Body Experiences in the Blind,* Kenneth Ring and Sharon Cooper investigate the evidence from out of body and near death experiences that those who are physically blind, including blind from birth, experience accurate vision during near death and out of body episodes. It is clear that during NDE's and OBE's neither the physically sighted or the physically unsighted are seeing with physical eyes. Ring and Cooper call this kind of transcen-

---

[258] Carter, C. 2010. *Science and the Near Death Experience.* Inner Traditions; Amazon review by Trond Skaftnesmo; text in square brackets added.
[259] Dossey, L. 1997. *Recovering the Soul: A Scientific and Spiritual Search.* Bantam.

dental seeing 'Mindsight'. A seeing providing accurate visual detail, sometimes from many angles at once, and one accompanied by synesthetic qualities of multisensory knowing. They note that:

> 'In general, blind people report the same kinds of visual impressions as sighted persons do in describing NDEs and OBEs. For example, ten of our twenty-one [blind] NDErs said they had some kind of vision of their physical body, and seven of our ten [blind] OBErs said likewise. Occasionally, there are other this-worldly perceptions as well, such as seeing a medical team at work on one's body or seeing various features of the room or surroundings where one's physical body was. Otherworldly perceptions abound also, and seem to take the form characteristic for transcendental NDEs of sighted persons – radiant light, otherworldly landscapes, angels or religious figures, deceased relatives, and so forth.'[260]

'This is yet another piece of potentially *epoch making* evidence that falsifies the materialist paradigm, a paradigm that has become a science blocker and knowledge stopper. Ring and Cooper situate the evidence in a multi–dimensional metaphysical matrix that accommodates these experiences, noting the parallels between modern western understandings of the quantum nature of consciousness and the eastern metaphysical traditions such as that of the yogis.

'There is, now, Ali, decades worth of research into Life after death, communication with those passed on, out of body experiences, near death experiences and into the phenomenon of reincarnation.[261] This accumulated research shows that these phenomena are real, that there really are multi–dimensions and vibrations of consciousness where those 'passed on' continue to evolve, in world's composed of 'matter' at higher vibrations than the 'matter' we know, (itself an illusion or maya 'made' out of nothing but force, frequency, space and vibration), including the possibility for returning to earth and the 'lower' material vibrations, for further Soul expression and Soul learning in facing the kinds of spiritual growth challenges which can be met with in these denser conditions.[262]

---

[260] Ring, K. & S. Cooper. 2008. *Mindsight: Near-Death and Out-of-Body Experiences in the Blind* iUniverse; p. 75.

[261] Stevenson, I. 1980. *Twenty Cases Suggestive of Reincarnation: Second Edition, Revised and Enlarged.* University of Virginia Press. Stemman, R. 2012. *The Big Book of Reincarnation: Examining the Evidence that We Have All Lived Before.* Hierophant Publishing.

[262] Newton, M. 1994. *Journey of Souls: Case Studies of Life Between Lives.* Lewellyn Publications. Over a thirty-year period Newton hypnotically regressed 7,000 clients,

'But these findings of modern spiritual–science, conducted by researchers at bodies like the British and American Societies for Psychical Research and the International Association for Near Death Studies or the Afterlife experiments of Dr Schwartz, the medically focused researches of neuro-psychiatrist Dr Peter Fenwick and cardiologist Pim van Lommel, demonstrating that out of body, near death and Life-after–Life phenomena are real have, in the past, generally been ignored because *they refute materialism.*'

'But surely once a scientific hypothesis is refuted it's refuted?'

'I agree, Ali, it is, but it's hard for many of those who have been lifelong supporters of the materialist cause to accept this. No one likes to admit that they might have been mistaken in their opinions, even if, by revising their views in the light of modern evidence allied to ancient wisdom, it means existence suddenly becomes so much vaster, richer, broader – so much more intelligent and so much more interesting to explore.

'In a Subjective and intelligent Cosmos or a meaningful Mystery of Existence, with a 'heads' or Consciousness side, there are higher dimensions, frequencies and planes of existence and consciousness to explore, in addition to the ones we are currently and most immediately familiar with. Just as we explore the physical, 'outer' world using our physical bodies, so it is possible to explore and connect with these parallel dimensions and planes, even during our physical plane lives, if we wish, by means our own inner beings and subtle, Soul bodies as our vehicles for travel in these dimensions – as the world's yogis and sufis, shamans and saints and sages, psychics and mystics, prophets and mages have always known.[263]

In *Less Incomplete: A Guide to Experiencing the Human Condition Beyond the Physical Body,* Sandie Gustus asks:

> 'How many of us have had an experience that suggests a deeper, unseen reality . . .? Most of us have probably had an experience with

---

building up a picture of aspects of the life between lives time as his clients, while in hypnotic regression, reported on the same themes again and again: including self-assessment of the concluded lives, further experiences in the spiritual dimensions before eventual return to physical life for continued Soul learning and growth.

[263] Peake, A. *The Out of Body Experience: The History and Science of Astral Travel.* Watkins Publishing. Monroe, R. A. 1972. *Journeys Out of the Body.* Souvenir Press. Yogananda, P. 1946. *Autobiography of a Yogi.* Self-Realization Fellowship. Gustus, S. 2011. *Less Incomplete: A Guide to Experiencing the Human Condition Beyond the Physical Body.* O Books.

déjà vu, intuition, synchronicity, premonition or telepathy; felt an instant sense of recognition or familiarity with a complete stranger … Such indications that our reality is more complex than it appears to be and that there is a kind of 'backstage' to our existence, are commonplace.

'Despite this, most people have no direct experience that we live in a multidimensional environment that extends far beyond the boundaries of our physical world, and that we are, in fact, much more than just our physical bodies.

'Fortunately, there is one phenomenon that is natural to all humans that allows us to verify for ourselves, from first-hand experience, that we are capable of acting entirely independently of the physical body in a nonphysical dimension… The out of body experience (OBE). Anyone who has had a fully lucid OBE, and I count myself among them, will tell you that if you can be lucid outside the body, you will find that all your mental faculties are fully functioning, that you can make decisions, exercise your free will, access your memory, think with a level of clarity that sometimes exceeds your usual capacity, and *even capture information from the physical dimension that can later be corroborated* . . . and that to experience all this provides you with irrefutable evidence that the physical body is merely a temporary 'house' through which your consciousness (i.e. your soul or spirit) manifests in the physical world.[264]

'Gustus's book is based on her own experiences and the research of Waldo Vieira which includes a great number of out of body experiences collected during the course of his life. She includes practical methods for inducing OBEs and for regaining our contact with the nonphysical realms of existence. This, obviously, has tremendous implications for our culture and, ultimately, our sciences. Gustus explores post 'death' or Life after Life reality, multi-dimensionality, past lives, the evolution of consciousness, multi–dimensional ethics and karma and makes interesting discriminations as to the differences between lucid dreams, OBE's and remote viewing.

'In *The Out of Body Experience: The History and Science of Astral Travel,* a book which 'attempts a short explanation of why quantum physics may be the unlikely source of answers to the mystery of the OBE experience,' Anthony Peake describes how since the publication of his first book, *Is There Life After Death? – The Extraordinary*

---

[264] Gustus, S. 2011. *Less Incomplete: A Guide to Experiencing the Human Condition Beyond the Physical Body.* O Books.

Science of What Happens When We Die he has met many people who have experienced OBEs firsthand. He says:

> 'I am regularly struck by just how down-to-earth and 'normal' these [OBE] experiencers are. They are ordinary people who have had extraordinary events take place for which they have no explanation. Indeed, many confide in me because they are scared that by telling their friends and associates of their experiences they will be branded as needing psychiatric treatment.

'Peake continues:

> '... science may be at the start of a new paradigm that can accommodate such anomalies... Indeed, many open-minded scientists regularly observe that the next catalyst will be a breakthrough in our understanding of the nature of consciousness. This is because consciousness, and its implications, present *insurmountable problems* with regard to the present [materialist] scientific paradigm. Self-awareness within *a seemingly unaware universe simply does not make sense,* and psychic phenomena similarly present anomalies that may be the pointers to a hitherto unknown model of science; *a model in which mind creates matter rather than matter creates mind.*[265]

We finished our drinks, got up and moved on, continuing up the valley, with the enchanting river by our side.

---

[265] Peake, A. *The Out of Body Experience: The History and Science of Astral Travel.* Watkins Publishing; emphasis, text in square brackets added.

15 Magical Living World: Mindless Mystery or Blazing Mystery?

What kind of a mysterious World are we in? Is it a random mystery as materialism believes or an intelligent mystery as Idealism maintains?

'The dogma of the materialist thinking that controls mainstream science today is that 'only-matter-exists' – and, therefore, *Consciousness, Life, Mind, Soul and Intelligence* can none of them be self-existing principles or, 'higher' Natural phenomena. As this is so clearly false the troubling implications for world science and culture are that materialism not only cannot account for the data of reality but also acts as a profound break on human progress. It blocks the expansion of our knowledge of the true character of total Holistic Nature and of our own true Nature, our capacity to understand Consciousness, Sentiency and Mind, the Life Forces and the soul in all things not to mention our progress and onward journeys once we leave this world.

'Materialism tries to make total Holistic Reality small, meaningless, mono-dimensional and unintelligent – and it fails. When it comes to our understanding of the ultimate sources and causes of our Living World the overwhelming evidence is that 'accidents' do not create 'clever machinery' or even simple 'machinery' of any kind, let alone CODES and software of the most stunningly sophisticated sorts to run that 24/7–Living 'bio-machinery'.

'Regardless of our labels for these ultimate sources and causes, be they 'super' Nature, the intelligently causal multi-dimensions, God, the Tao, Buddha Nature or Supreme Intelligent Beingness, it is clear that Living Nature and our minds and intelligences are all sourced in amazing intelligence and purpose not mindlessness and accident.

'This has been the near universal view of humanity, the classical, idealist view which says that Consciousness, Intelligence and Mind are all *ontological* qualities, they constitute the warp and the weft, the very *fabric of existence* and they give rise to the apparencies or illusions of matter and materiality which, as the yogis pointed out centuries ago, and as modern quantum physics has confirmed, are not ultimately, truly 'solid' or 'material' at all but are made entirely of clever vortices of forces and energies, that is of intelligently in-form-ed, quantum–energy s—p—a—c—e.

'No one can say *why* the qualities of Consciousness, Intelligence and Motive Forces exist but only that they do. The classical, idealist view is that they are *senior to and prior to* matter and manifestation – they are its causes; they give rise to it. The impossible thing is Existence – for why should Existence exist? Anything at all rather than nothing at all? All anyone can ask is which view makes better sense of the data of reality? *Which view is the better scientific hypothesis?*:

(a) That self-existing, self-arising Mindless, Unintelligent, Insentient Matter plus trillions of Accidents should *'just exist'*, who knows why? This is the hypothesis of materialism: The 'mindlessness-unintelligence-and-accidents-before-everything-intelligent-and-clever-in-Nature' view of Existence. Or,

(b) That Self-Existing, Self-Arising Consciousness–Intelligence–Love–Purpose–Matter should *'just exist'*, who knows why? This is the hypothesis of idealism: The classical 'mind-and-intelligence-before-mechanism' view of Nature.

'If modern physics in combination with humanity's cumulative spiritual experience is allowed to be our guide, then the matter is already decided:

There is no matter as such. All matter originates and exists only by virtue of *a force* which brings the particle of an atom to vibration ... [For Max Planck such force implied] *conscious and intelligent mind.*'[266]

'Sir James Jeans, another eminent physicist and idealist also noted that, in the light of the discoveries of modern quantum physics:

'... the Universe begins to look more like a great thought than like a great machine. *Mind* no longer appears to be an accidental intruder into the realm of matter... we ought rather hail it as the creator and governor of the realm of matter.[267]

---

[266] *Das Wesen der Materie* [The Nature of Matter], speech at Florence, Italy (1944) (from Archiv zur Geschichte der Max-Planck-Gesellschaft, Abt. Va, Rep. 11 Planck, Nr. 1797). Emphasis added.

[267] J. Jeans, *The Mysterious Universe,* Cambridge University Press, 1931, p.137.

'And Antony Peake:

> '[Human] Self-awareness within *a seemingly unaware universe simply does not make sense,* [this points to a model of reality] in which *mind creates matter rather than matter creates mind.*'[268]

'These views of modern western thinkers confirm what the advanced Yogi-Adepts of India said centuries ago, that the Cosmos is, ultimately, a mental emanation, like a great Cosmic dream.'

'I like that, Ollie. It makes me think of what the aborigine people of Australia say – that "there's a dream dreaming us."'

'Yes, in a sense Life is but a dream. But is it a dumb dream or an intelligent one! The ideological–materialism that so soul and meaning destroyingly controls science and biology today teaches the world's children and young people that existence is *intrinsically unintelligent, inherently purposeless and essentially meaningless.* Philosophical Idealism, by contrast, takes the view that existence, consciousness, mind, intelligence, the universe, Living Nature all make more sense when understood as the purposeful expressions of a blazingly Intelligent Mystery of self-existing and self-arising *Consciousness and Intelligence* that is prior to and itself gives rise to the illusion of 'solid' m—a—t—t—e—r 'made' out of intelligently informed quantum–energy s—p—a—c—e.

'But doesn't 'intelligence' imply 'purpose', which implies intention and motive?'

'Yes.'

'So we can't stop at simply saying that: 'The universe is *intelligent*, now carry on living as normal.' Because surely, then, we want to know if that intelligence is 'good'? Does it wish us well? Is it personal or impersonal, do we have a relationship with it or to it and so on?'

'Yes, those are all good questions – but they are not ones that physical–science can answer. Those are questions that need to be addressed to the world's religious and spiritual traditions, not physical–scientists. Physical–science *can* help us, however, to tell the difference between *intelligent* mechanisms and *'accidents'* within the Natural and the Living Worlds. Nothing should cause us to believe a Jaguar Fighter Jet is an 'accident' or that even just one tiny Living Cell, which is *many times more complex* than a Jaguar Fighter Jet, could possibly be an 'accident' either.

---

[268] Peake, A. *The Out of Body Experience: The History and Science of Astral Travel.* Watkins Publishing.

'That knowledge though doesn't tell us about the moral worth or purposes of the intelligent mechanisms we are studying. Is a cleverly made Jaguar Fighter Jet a 'good' thing or a 'bad' one? If it's shooting at us we'll say it's a 'bad' one, if it's protecting us we'll say it's a 'good' one. Is a highly intelligent, Jaguar–Cat–Soul–Being a 'good' thing? It kills other animals to live. Maybe that's 'bad'. Yet, perhaps it helps to keep its prey-species fit, weeding out the weaker members; so maybe that's 'good'.

'As to the nature of *the overall intelligence* within which we live and move and have our being – one which gives rise to Living Cells, Fish, Amphibians, Dinosaurs, Birds, Insects, Mammals, Plants, Trees and People, and from which our own amazing *intelligence* ontologically derives – we look to the spiritual–sciences, the world's religions and spiritual traditions, the esoteric teachings, like cabala, theosophy and anthroposophy, the yogic teachings and modern psychical and spiritual research to make sense of it. We don't look to physical–science, which is, fundamentally, concerned with *matter* and *mechanism,* not with consciousness, subjectivity, sentiency, intelligence, *purpose* and *meaning*.

'Physical–science only went wrong, in recent times, when it strayed into the territory of spiritual–science and began to tell the spiritual–scientists (the world's religious, spiritual, esoteric and psychical school students and teachers) that what they were studying, their magisterium, as the late S. J. Gould referred to it – the ultimate meanings and purposes of things, humanity's *higher* evolution and *higher* purposes – was all just so much 'make believe' and the study of non-existent phenomena ... that there *were* no Life or Vital Forces, that there *was* no Soul, there *were* no higher or spiritual multi–dimensions beyond or behind the visible, outer, third dimension of physical Nature, that there *was* no Life after death, that there *was* no intelligent, *cosmic scale evolutionary plan* unfolding and so on. Existence was all just one big 'accident'.'

'But you're saying that we can be Idealist – believe that the Universe, and our Living World, are Sourced in *Intelligence* or have *intelligent* evolutionary causes without being religious?'

'Yes, exactly.'

'What about you Ollie, do you have a religious view?'

'I don't belong to any religion, as such, Ali. I think there are deeper and higher meanings to existence, and that we can connect with these. I do believe that our universal witnessing awareness, our sentient, conscious minds and intelligences arise out of a Universal Consciousness and sentiency, *a blazingly creative, evolutionary intel-*

ligence that has many dimensions and vibrations and that the more we consciously and directly connect with that universal energy and intelligence – to which humanity gives blanket names like God, Brahman, Allah, Buddha Nature, the Tao, Supreme Intelligent Being or simply the Great Mystery – the better, the more beautifully, the more truthfully and kindly we can Live.

'That, however, takes inner work, meditation, mindfulness visualization and prayer. Humans are beings in consciousness and spiritual evolution far more than they are in physical evolution. Our physical bodies were perfected long ago. The evolution we are embarked on now is one of consciousness, of, among other things, marrying head to heart so that our global civilization is informed by love and fellow feeling more than by selfishness and excessive competition.[269] We are evolving from self-centeredness to all-centeredness while, at the same time, retaining our sense of individuality.

'That's why I care about this subject. I do not believe that we can best connect to the universal intelligence, the universal Life Source and wisdom stream, that is all and informs all, by solely looking outwards in our telescopes into outer space or downwards with our microscopes and particle colliders into sub-atomic space, useful and clever as those activities are, but only by directing our gaze, awareness and presence inwards into spiritual and inner space, by meditational and other practices (including psychic and higher-sensory development) where our vehicles for travel are our mental, emotional and Soul bodies, not the physical body and physical senses with which we explore physical space.

'The western world has, for too long, over emphasized the vita activa and underemphasized the vita contemplativa. This has contributed to the extreme materialism and its accompanying de-intelligenced, de-Souled, de-purposed perception of a Cosmos of mindlessness, unintelligence and 'accident'. As our Soul Life dried up we increasingly projected our own dessicated, Faustian Soul-emptiness ever outwards onto Living Nature and saw only 24/7 mindlessness, unintelligence and 'accident' wherever we looked, when, actually, Nature's Living, Soulful reality was the precise opposite of what we and our science increasingly claimed to be seeing.[270]

'By way of traditional religion the yogic or Hindu tradition makes a lot of sense to me because it is broad minded, generally non dogmatic

---

[269] Bucke, R. M. 1905. *Cosmic Consciousness: A Study in the Evolution of the Human Mind.* Innes and Sons.

[270] Tompkins, P. 1997. *The Secret Life of Nature.* Harper Collins.

and because it teaches, at its highest levels, beyond the plethora of divine energies that it recognizes, that All is One, that nothing exists but God, Brahman or supreme intelligent Beingness – a universal, intelligent-energy-being-essence who or which is both within and beyond the eons long Creation-in-Evolution or ELCIE, a supremely intelligent Beingness with which it is possible to have the most personal and intimate of relationships, (divine love, mystical union) yet who and which is, at the same time, beyond all knowing. Within this tradition there is the goal of god-realization or enlightenment (perhaps the closest concept in Christian terms would be to be 'truly born again in spirit'), in which we can come to fully realize that we are not separate from the all pervading divine energy, that, ultimately, it is what we are 'made of', we are one with it – this the Hindus and Buddhists believe brings enlightenment and great inner peace and joy.

'The difference of the eastern concepts of god–realization, self–realization or enlightenment with the Christian concept of spiritual rebirth and of salvation is that the Soul's curriculum lasts, in the great majority of cases, for more than one human Life and it cannot be fulfilled, comforting as that thought is, simply by professing certain beliefs and creeds but only by a gradual movement from self-centeredness to all-centeredness and a realization of the true Divinity, Oneness and holiness of all that is, even while in a physical body, because there is nothing that is not part of supreme Being, there is nothing that is not, ultimately, part of the All, that does not come from it or return to it.

'The Hindus also recognize that for manifested Existence to be, there has to be *duality*, that is good and bad, hot and cold, up and down, left and right, gradients and relativities of all kinds. And, as this is the case, they see Supreme Intelligent Beingness as the One, Brahman, that becomes three, a Creating Energy, Brahma, a Maintaining One, Vishnu and a Destroying or Recycling One, Shiva.

'The yogis aver that behind all the challenges of outer existence there beats a heart of blazing Intelligence, amazing Love and potent Purpose, and that Humans have the sacred possibility, even during their physical lives, to make a true and fulfilling connection with that Source or Supreme Nature and to make the magical discovery that, ultimately, it is their *own* True Nature too, that they really are 'made in its image' and partake in its qualities of Intelligence, Love and Purpose, Beauty, Goodness and Truth, Sat–Chit–Ananda [knowledge-consciousness-bliss]. The evolutionary journey to the full realization of that discovery or to *moksha*, as the Yogis and Hindus say, the progressed state that the Buddhists call enlightenment or nirvana, and corresponding release from the cycle of death and rebirth, is considered to be a long journey on which every human Soul is embarked, whether they are consciously aware of it or not.

'The materialist view, that controls science today, is in conflict with and opposed to all this: It says 'accidental', 'unintelligent' matter precedes Consciousness, Sentiency and the spiritual essences of Intelligence, Love and Purpose, Goodness, Beauty and Truth, which are all 'accidents'. In materialism's de-Souling scheme these spiritual essences and Soul qualities have no ontological existence. They are not eternally existing verities, they are simply 'accidental' mental constructs. Materialism believes that solid, stolid, unintelligent, mindless, numb, dumb, tiny billiard ball atomic matter 'accidentally' creates Consciousness, Awareness, Sentiency, Intelligence, Mind, miraculous Life and magically and Soulfully Living Beings.

'*The materialist hypothesis,* which, as we've seen, depends on completely ignoring and actively rubbishing vast swaths of human experience and knowledge, is a very simple, simplistic and unreasonable view of Life and existence. It ignores our universal experience that intelligent mechanisms and clever machines – the Soul Life Form Beings being the cleverest 'mechanisms' or 'machines' known to humanity – *only ever* have intelligent and clever sources, they never arise or evolve 'by accident' – the forces of entropy alone prevent such a thing. It also dismissively discounts humanity's diverse spiritual traditions and their insights into total Reality. When it comes to political power, as it did in Soviet Russia, Cambodia and elsewhere, it even abolishes them and makes them illegal. The materialism that has captured science and biology and falsely declared that 'science *is* materialism' is not only unreasonable but it is a dismal and nihilistic creed.

'Followers of materialism have to irrationally regard even their *own* minds and intelligences as the accidental and purposeless byproducts of unconscious, dead, zombie atoms which they also, of course, see as purposeless and accidental.'

'But that's not logical …'

'I agree, it isn't, but materialism 'has' to think this way, because if it concedes that even very matter itself might, ironically, be extremely smart, then its thought system collapses because it is based on the dogma that existence, including matter, is *unintelligent* and without purpose or teleology. Whereas the Idealist scientific hypothesis has no difficulty with the concept of very matter itself being 'smart' and of having many meaningful purposes and functions.'

'*Intelligence* is an ontological and spiritual *essence*. It is part of the self-existing fabric of total Reality. We have no reason, let alone any evidence, to think that *intelligence* arises from atoms of carbon, oxygen, hydrogen and nitrogen bumping into each other by Darwinian 'chance', accidentally becoming bacteria shaped and by another trillion 'accidents' eventually becoming people shaped, at which point, voila, *'intelligence'*. The very primordial smartness of those clever atoms, each a tiny mechanism in its own right, is an expression of the *same* amazing, self-arising, self-existing, *pristine intelligence* that our own minds and intelligences arise from.

'*Intelligence* which we recognize in ourselves as an *inherent* ability to respond to the world with awareness, cleverness, learning, knowledge, comparison, evaluation, instinct and intuition, a characteristic shared not only by humans but animals and plants[271] is generally considered to be, in this materialistic age, a product of brain functioning rather than a fundamental quality of existence, of Holistic Nature, of all that is. However, in *Brilliancy: The Essence of Intelligence,* A.H. Almaas, author of *The Unfolding Now: Realizing Your True Nature Through the Practice of Presence,* explores the ontological nature of *intelligence* as a fundamental quality of consciousness and of the ground-state-beingness of existence. He explains how it is possible to experience the *essence of intelligence* directly, as a self-arising quality of consciousness that he calls *brilliancy,* rather than by the usual means where we see our intelligence second hand, through the observation of its manifestations, for example in the quality of our thinking and actions. He says:

> 'The understanding reflected in this book is a result of a particular spiritual transformation that *reveals the ground and nature of consciousness.* This ground turns out to be the underlying nature of everything, even the physical universe – and hence the body and its brain. This spiritual ground, what we call Essence – the essence of consciousness and all of reality – reveals itself through many qualities, which are primordially inherent to it. ...
>
> 'One thing we discover in this revelation of the nature of spirit is that it is characterized not only by qualities such as power, love, truth, and so on, but also by a particular *luminosity* that appears to our mind to be *intelligence.* In other words, we realize that *we can actually experience intelligence directly* – not through an activity, as we normally do, but as *a palpable presence,* as a presence of pure consciousness characterized by *intelligence.*

---

[271] Tompkins, P & C. Bird. 1989. *The Secret Life of Plants.* Harper Perennial.

'We find out that *intelligence* is an inherent quality of our spiritual nature, fundamentally inseparable from it. Yet in functional activities, it flows through our consciousness, and through its physiological supports – the brain and the nervous system – to give these functions a kind of efficiency and completeness we usually associate with intelligence.

'We discover our spirit *as* intelligence, intelligence that manifests in what we call intelligent functioning, yet is *experienceable directly* and apart from its functioning.

'We experience ourselves, then, not as *having* intelligence but present *as* intelligence. In this experience, we find out what the *essence of intelligence* is, the source of this capacity.'[272]

'Almaas helps us to understand that *intelligence* is a fundamental and self-existing essence. It is part of the fabric of reality, as many Near Death Experiencers and spiritually realized people have said, and we, as humans, through practices of inner inquiry and meditation can experience this *essence of intelligence* directly, unmediated by our thoughts or actions. Atwater describes NDErs who:

'felt awed by the wonderment of a blackness [or light] that appeared intelligent, emoted feelings and instilled in the experiencer a sense of peace and acceptance.[273]

'Almaas points out that the more access a human being has to the direct and unmediated essence of intelligence, that is part of the ground–state–beingness of reality, the more efficient, capable, elegant and intelligent their functioning. This also ties in with the eastern concepts of enlightenment and self-realization which lead to greatly enhanced personal functioning. This understanding of intelligence as a primal spiritual quality belies the materialist idea that *intelligence* is a product of the brain but rather makes it clear that the brain, as indeed all of Nature, is a manifestation and production of *intelligence*.

'This is why we can say that tiny Living Cells, *irreducibly* complex, there from the very beginning, more than 3 billion years ago, stunningly complex from the first, with no evidence anywhere for a Darwinian-style movement to gradually rising complexity in their emergence, are expressions of incredible *intelligence*. This is why we can ask:

---

[272] Almaas, A. H. 2006. *Brilliancy: The Essence of Intelligence*. Shambhala; italics added.
[273] Atwater, P. M. H. 1994. *Beyond the Light: Near Death Experiences*. Thorsons; p. 28.

'Is it really true that the scientific hypothesis of materialism still stands or has it been refuted? Is it really true that the scientific hypotheses of mindless, unintelligent and accidental causation of Life and random, unintelligent evolution of the Living World are correct? There is certainly no evidence for them. Is really true that materialism's knowledge-gap-filling, pseudo-uber-creative-gods of unconsciousness, unintelligence and accidental chemistry are responsible for the astonishingly complex bio-chemical processes going on 24/7, literally trillions, every second in every human body, all day long, all Life long?'

'I don't think so, Ollie,' Alisha said, smiling at my passion.

'This is why I'm so frustrated with materialism's takeover of classical science and biology, Alisha, for which I have huge respect. You see not only is Soul-and-intelligence-in-Nature-denying materialism a vast delusion or collusion, it also actively rubbishes *all* the higher truths, religious insights and spiritual meanings that humanity holds so precious and, as Max Planck said, "has for the pious belief in a higher Power nothing but words of mockery.".'[274]

'But isn't that as much a reaction to simplistic and outdated religious views as it is to anything else?'

'That, Ali, is not a good enough reason to deny the obvious *intelligence* and purposefulness of the incredibly smart mechanisms of our 24/7–dynamic Living World. There are middle ways you know. Such thinkers could say:

> 'We don't believe in simplistic, bearded-god-created notions of an intelligent multi-dimensional, multi-vibrational Cosmos, but it is quite clear that we exist as part of a multi-frequency Living Nature that is *incredibly intelligent, not its very opposite,* and it's fabulous to explore it.'

'In other words, Ollie, the philosophical materialists could convert to philosophical Idealism?'

'Yes, they could. The great thing is, Ali, these thinkers could drop the 'materialism' from their science without the need to join any religion or even any need to agree that the *intelligence* and *purpose* that pervade the Living World are 'good'. They could even argue, if they wished, with their philosophical hats on, and in agreement with Plato and the Gnostics, that they are *imperfect and flawed.*

---

[274] Planck, M. 1958. *Religion und Naturwissenschaft*.

'Many people would argue a Stealth Bomber is an 'imperfect' and 'flawed' device, morally speaking, but no one denies the lethal *intelligence* that it embodies and the beauty of its form. We may not like Scorpions or Boa Constrictors – but deny their intelligence and beauty? Pretend that we're looking at Darwinian *'accidental chemical reactions'* when we observe them? Is that *really* how chemicals behave in the real-world? Is there *any ordinary evidence, let alone extraordinary evidence anywhere for the unreasonable claim* that this is how the unguided, Life-Forceless, Soulless chemicals of materialist faith and belief actually behave? They 'accidentally' become Living, Breathing, Breeding Cells more complex than fully-sized chemical factories? They accidentally become the astonishing, Life Forms of the magical and sacred Soul Beings of the Plants, Animals and People? Tumble chemicals in barrels and out pop COBOL CODED computers or DNA CODED Life–Form–Soul Beings vastly more complex than COBOL CODED computers?

'Darwinian unintelligent design and materialist thinking do seem to be unreasonable ways of looking at reality, Ollie.'

'That's the point, they are unreasonable. Materialism is unreasonable. It is a distortion, a corrosive and Soul-damaging lie that the world's children and young people are being exposed to everyday. What could be more unreasonable than the materialist dogma that *the only mechanism for all existence, all* being, *all* actuality, *all* potential, *all* movement, *all* Life, *all* mind, *all* desire, *all* intelligence, *all* purpose, *all* change, *all* evolution in Nature, be it the existence of mysterious gravity, exciting electricity, fabulous light, magical magnetism, the beautiful elegance of $E=mc^2$, astounding atoms, amazingly smart-materials chemical molecules, DNA CODE ("more advanced than any software ever created"), tiny Living Cells ("the most complex systems known to man"), the stupendously complex multi-cellular Life Forms, the Ants, Lions, Oak trees, Hawthorns, Rabbits, Otters, Badgers, Butter Cups, Rainbow Trout and Heron, the brilliant minds of Newton, Bach and Einstein is potential-less, mindless, powerless, insentient, un-constructive, unintelligent *"accident"!*

'You mean a mindless and unintelligent Cosmic Mystery of existence? Yet what is the evidence for that materialist view point?'

'There isn't any. Unless the fact that there are no bearded-gods anywhere to be seen counts as evidence. Materialism, in science, is simply a non-evidence-based belief system, a kind of closed-minded pseudo–religion that some people call scientism – I sometimes think of it as 'zcience' for it is completely topsy turvy and back to front.

There is no evidence anywhere that 'accidents' are evolutionary except when it comes to relatively unintelligent features in Nature like randomly evolved landscapes. 'Accidents' cannot account either for Life's origins or for its clever CODES or for its intelligent evolution through vast time or for its second by second, 24/7–clever functioning right now.'

'And, it is clear that, without inherent, self-existing, self-arising urge and stupendous potential, potent purpose, flaming desire and blazing cosmic scale *intelligence* there would be nothing to disturb the absolute silence of absolute nothingness, there would not be the movement or existence of even one quark. Yet, as soon as there is departure from absolute stillness, there is movement, which is doing, which is acting, which is function, which is purpose, which is intelligence. This is basic ontology. Without desire or urge there is no departure from absolute stillness and nothing would ever be. Even 'accidents' can only happen within Being, never outside it. Being that contains potentials which imply intelligence, desire and, ultimately, stupendous power, enough to give rise to a universe that is billions of light years across. Such an extra-ordinary and highly 'super' thing does not imply 'numbness, dumbness and unintelligence'. It implies blazing, dazzling, vaporizing power and brilliance. Materialist thought, however, will have none of this and purposeful, meaningful, *intelligent causation* and *intelligent* evolution, in the Living World, is forbidden.'

'But that seems so false to reality ...'

'Ali, what do you think ideological materialism is? Its unreasonable, non-evidence-based, knowledge–gap filling 'explanation' for everything is 'accident', always and everywhere, all the way up and all the way down. This is why this school of thought so dislikes the hypothesis of intelligent design and intelligent evolution in biology or the large body of modern evidence for extrasensory abilities and for Life after death because these diverse pieces of evidence all confirm the intelligence and purposefulness of the multi-dimensional Cosmos in which we live and move and have our being. A conclusion which materialist thought, oddly, seems to find depressing rather than stimulating and exciting.

'One thing, though, Ollie, given that you don't belong to a specific religion I'm surprised that you care so much about these issues?'

'Because what is *actually* true is not only interesting, it really matters. Is the evolution of Living Nature, across vast time, a random process or an *intelligent* one and are scientists capable of finding out? If it's the latter what does it all mean? Is it good or bad or merely indifferent? Are the amazing Soul–Life–Form Beings, of the Plants, Animals and People 'accidents', or are they part of a mysterious but *intelligent* evolutionary scheme? Which of these questions can material–science (biochemistry, biology, etc) answer, and which of them can spiritual–science seek to answer? That is all those disciplines exploring into the subjective dimensions, vibrations and frequencies of Existence – beyond the purely material and physical.

'The world's mystical, hermetic and esoteric traditions, like the teachings of the yogis, the sufis, cabala, theosophy and anthroposophy are all important elements in an emerging, modern, spiritual–science, of which contemporary psychical, Near Death and OBE research are also all parts. The interesting thing about the traditional and more modern esoteric and psychical research, for our conversation about evolution, Ali, is that they all maintain that it is the *Soul* and *Spirit* in the Life Forms that is as much in *intelligent* eons-long evolution as the outer Life Forms themselves which are considered to be but the outer vehicles for the total Cosmic Soul's exploration of all facets of existence, the chemical, the mineral, the plant, animal, human and spiritual. The spiritual–science understanding is that the total Soul, the 'heads' side, the subjective and Spiritual side of the 'coin' of Holistic Nature, hologrammatically self-fragments into and extends Itself as all the dimensions and levels of existence in order to explore, experience, unfold and extend. The whole thing – to which people give overarching names like God, Brahman, the Tao, Allah, Buddha Nature or simply the Great Mystery – is a vast, self-arising, self-extending, self-creating cosmically scaled exploration of consciousness, sentiency, creativity, expressivity and inner, subjective growth.'

'But why does it exist at all?'

'No one can answer that. All we can do is decide whether the data of reality, of which our own Consciousnesses, Sentiency, Minds and Intelligences are all a part, point, more reasonably, to a mindless, unintelligent and accidental Cosmic Mystery of Existence, a meaningless MUA or to a blazing, amazing, ultimate Miracle of a Cosmic Mystery of Existence, a meaningful AUM.'

'Those are, really, our only two choices?'

'Yes. The trouble is our Western minds have lost touch with the true Life, Soul and Spirit of things. We've lost touch with the intelligent Cosmos and sacred Gaia, thinking she's just an insentient rock and that Life isn't really 'Life' at all, just some kind of world-scale *'accidental chemical reaction'*. In the unreasonable and impossible materialist ideology of mindlessness and accident these buttercups and dandelions and meadow grasses, the trout in this lovely river and the otters fishing for them, the cows in this field, the sheep on the hill, the glossy black crows, humanity, all sentient beings are all just 'purposeless', unconscious zombie' *'accidental chemical reactions'!* – if we cannot see how absurd and unreasonable that is it is a sad day for our western wisdom tradition.

'Is it really true that Living Cells, 'world[s] of supreme technology'[275]... 'veritable micro-miniaturized factor[ies] ... *far more complicated than any machinery built by man* and absolutely without parallel in the non-living world'[276] are good evidence for the mindlessness, unintelligence and 'accident' that materialist thinking science and biology believe are the purposeless and random sources of all that so amazingly is? Is it really true that human brains, each with more molecular-scale switches than *all the computers, routers and internet connections on the entire planet* are signs of mindlessness, unintelligence and accident? Where is the extraordinary evidence for these extraordinary claims? There is absolutely none.

'These phenomena don't indicate a lack of intelligence or 'teleology' in Nature. They call out purposefulness and intelligence to us. The trouble is that ever since the intellectual west pricked its finger on Darwin's pin, so many of us were sent to sleep, like Sleeping Beauty, for 150 long years now, into a topsy-turvy dream-world where existence became meaningless, where 'accidental chemistry' supposedly, and with *no evidence at all* to back up this unreasonable and anti-empirical claim, suddenly became entropy–defyingly 'creative' and 'evolutionary' and an allegedly, potential-less, impotent, numb, dumb, Soulless, Vital–Forceless Cosmos of utter mindlessness and unintelligence accidentally led to *intelligence* and brilliant Minds including those of Plato, Galileo, Bach, Newton and Einstein.

This de-humanizing, mineralizing thought system of a de-Souled Cosmos and a de-purposed Living World eventually began to deny the subjective components of Holistic Reality altogether even denying its own mind and intelligence, saying:

"We're all zombies. Nobody is conscious."[277]

'This is the mindless, unintelligent, and mono-dimensional cosmos of materialist thought where, surreally, intelligent minds begin to argue for their own unconsciousness, Soullessness and ontological unintelligence. The materialist idea (ideas are not material) is that we are astonishingly complex 'accidental' robots who are intelligent enough

---

[275] Denton, M. 1985. *Evolution: A Theory in Crisis.* Burnett Books; p. 328, emphasis added.
[276] Ibid. p.250, emphasis added.
[277] Dennett D. 1992. *Consciousness Explained.* Back Bay Books, p. 406.

to do science but that unlike robots in non-Darwinian reality we are 'robots' that 'mindlessness, unintelligence and accident' made!'

'This is when the amazing and magical offspring of multi-dimensional, multi-vibrational Divine Cosmic intelligence and meaning, blessed with free will, begin to use their very *immaterial* Souls, Minds and Intelligences to insist that they possess none of these. This is a tragic form of intellectual, emotional and spiritual self-harming that a science, distorted by the dogmatic faith system of materialism, has led us to – Consciousness denying Consciousness, Mind denying Mind, Intelligence denying Intelligence, Soul denying Soul, Spirit denying Spirit.

'But that is the meaning of 'free will', Ollie.'

'Yes, you're right. We are free to use our Minds as we wish – even to deny their own real existence.'

We paused for a moment admiring the beautiful view and the delightful river, before continuing along the valley.

'What we are disputing, though, is *not* the concept of evolution, in the sense of tremendous changes in the Living World, over vast time, but whether the mechanisms that drove those processes were *random* as materialist ('accidents-before-all') thinking holds, or *intelligent* as idealist ('intelligence-before-all') thinking and the world's religions and spiritual–science traditions maintain.

'Materialism makes the unreasonable claim that its potential-less, powerless pseudo-uber-gods, mindlessness, unintelligence and accident explain all. Yet, *can you think* of any examples of 'accidental' chemistry building or evolving anything intelligent? Can you think of 'accidents' of any kind building and maintaining *any* 'machines', even simple ones?'

'No, now you mention it, Ollie, I can't.'

'But that is the impossible and irrational foundation of sand on which the materialist accounts of the Living World's origins and unfoldment through time are both built, Alisha. Obviously, we don't make Cars, TVs and Computers by randomly throwing their parts around or tumbling them in barrels labeled 'Cars', 'TVs' and 'Computers' in the hope that, at the end of the tumbling process, out will pop shiny new Cars, TVs, Computers.

'Yet this is how materialist science and biology both continue to argue inspite of all the disconfirming evidence that the Living World arose and, as importantly, continues to arise right now, 24/7, blazing second by blazing second – as though, surreally, we're all living within some vast, meaningless never ending *'accidental chemical reaction'*. This is not a reasonable let alone an evidence-based world view that has infected the intellectuality and almost all the academies of the Western World since Darwin's pin put us all to sleep for the last 150 dreary Darwinian years.'

'And the forces of the materialist ideology have to think this way because …'

'Because they insist that we Live in a 'meaningless and accidental' universe and in conflict with almost all the greatest thinkers and scientists that have ever lived, in conflict with all of collective humanity's intuitions and mystical and spiritual experiences to the contrary, they maintain that there simply *are* no subtle, *intelligently* causal or spiritual multi-dimensions in a total Holistic Nature that *could* intelligently and meaningfully give rise to our Living World, magical, amazing Gaia and all its strange and wonderful Soul-Life-Form Beings, the Plants, Animals and People.

'We all had pictures of the modern Materialist Creation Myth in our text books in school deceptively implying, as a result of the famous Miller–Urey experiment, that Life was probably a 'chemical accident'.[278] What they didn't emphasize, in those textbooks, was that the experiment merely produced various amino acids, some of Life's basic building blocks.

'Miller–Urey, Ali, was like an experiment that was designed to show that Ancient Rome (the Living World) probably arose 'by random processes'. So they tumbled some clay and water in a barrel, heating it and when a few lumps of fired clay (representing amino acids in the metaphor) popped out that could just about be used as bricks in a Roman building (a Living Cell) they claimed that the semi-accidental production of these building blocks (the experiment was, after all, intelligently designed) proved that the buildings of Rome (the Life Forms) were 'accidents' because "Look, they too are made out of pieces of fired clay [amino acids]".'

'But a few pieces of randomly fired clay were not what the Capitol, the Forum or Rome, as a whole, were, Ollie.' said Ali, laughing.

'You're right. That Life could be an 'accident' is as unrealistic as claiming that a Jumbo Jet, far, far simpler than even just one Living Cell, is the incredibly *intelligent* result of random events of various kinds – high winds blew through the Jumbo Jet assembly plant and that was how it came to be. And, as Dembski says:

'The odds of the early Earth's chemical soup randomly burping up [a Living Cell] are *unimaginably longer* than 1 chance in 10 to 150.

---

[278] Wells, J. 2002. *Icons of Evolution*. Regnery Publishing.

[the universal probability boundary beyond which nothing can conceivably arise by 'chance' alone]'[279]

'The trouble is, Ali, the materialist view of Living Nature and what it is, a billions year long running *'accidental-chemical-reaction'*, is not only scientifically impossible, it's also socially, emotionally and spiritually corrosive.'

'Because ...'

'Well, firstly, it's *just not true* and science is supposed to be about what's true, it's supposed to be evidence-based – it's not mean't to be an unreasonable quasi religion that worships 'mindless-ness, unintelligence and accident' as Life's impotent and impossible supreme causes – and with no evidence at all but its own beliefs to back its extraordinary faith claims. The classical philosophers talked of the 'Good, the Beautiful and the True.' Materialism is none of these. It de-purposes and de-Souls the Living World. It denies the intelligent, multi-dimensional nature of reality, it denies the modern evidence for the survival of consciousness beyond 'death'. It teaches the world's children and young people the materialist line, be it a delusion or a collusion, that Living Nature's incredibly dynamic, 24/7–clever Life and Soul Force driven bio-complexities are the *amazingly intelligent results* of utterly mindless and chance processes. What could be more dispiriting and Soul destroying?'

'I wish we could do something, Oliver. We have to change this way of looking at Life and the Living World, it wants to reduce Living Nature to 'dead' physics,[280] it destroys all Soul and meaning.'

'I passionately agree because the modern materialist distortion of science that regards Living Nature as Life-Forceless, Soulless, 'purposeless', 'unintelligent' and 'accidental' matters. Firstly, it makes science irrational and misleading in our culture as to the most important matters of all, what we really are, the meanings and purposes of our existence. We are children of intelligence, Soul and meaning not of mindlessness, unintelligence and accident. Telling us we are all Life-Forceless, Soulless, zombie 'chemical accidents' is hardly inconsequential, especially as it is such patent nonsense. It has caused untold psychological and emotional suffering for generations now.

---

[279] Dembski, W and J. Witt. 2010. *Intelligent Design Uncensored*. IVP, p. 71.
[280] Crick F. 2004. *Of Molecules and Men*. Prometheus, p. 10. "The ultimate aim of the modern movement in biology is to explain all biology in terms of physics and chemistry."

I looked at Alisha, her kind face looking sad. I said, 'Don't despair, Ali. Thankfully, it is easy to show and some scientists already have, as we saw with Denton, Dembski, Behe and Meyer's work, that the materialist myth that Life is an 'accident' is either a widespread but sincere hallucination or it's just a collusion, in parts of the scientific community, to keep the priests and theologians at bay.

Alisha still looked sad as she reflected on the enormity of the damage done to the World's children and young people, and to our relationship to the Living World and all Sentient Beings by the materialist thought system.

'Materialist thought does, however, have just one small merit ...'

'What, Oliver? How can it possibly have any? It is so Soul destroying.'

'I did say it was only a small merit.'

'OK, so what is it?' said my friend, calming down a little.

'It is that it can free us from various kinds of outworn religious and intellectual baggage, giving us some space to try to create, hopefully, a better world.'

'OK, that is one positive. It's true that the world's religions can be very dogmatic and narrow-minded at times.'

'Exactly. But the trouble is that because it leaves us free to create a supposedly brave, new and better world *without reference* to anything higher than our biology, and personal and collective egos, it immediately becomes *dark*. This is why, ultimately, it doesn't work.

'The world's *spiritual*–science or esoteric traditions teach, on the other hand, that Humans are magical and intelligently evolving Soul beings in an eons long *intelligent* evolution. That we are in a vast, long-term process, as intelligently (not 'accidentally') evolving Humanity. That together, as a group or family, we are moving from primitive self-centeredness towards a special and precious condition of selfless inner and outer spiritualization, while retaining the gains of our experiences and learnings along the way. We make this journey, some more rapidly, some more slowly, over many human Lifetimes. This is where reincarnation comes in as it permits the gradual and fulfilling human Soul movement from self-centeredness to all-centeredness.

'Materialism, however, believes that there *is* no spiritual potential in us *higher* than our allegedly accidental biology, 'selfish' gene animality and our egos. Marx, Lenin, Stalin and Hitler were all Darwinian–materialists and they tried to reconstruct the world along so-called scientific–materialist lines. For example Hitler decided to put Darwin's ideas about the 'survival of the fittest' into action. Remember the full title of Darwin's work?'

'No, what was it?'

'It was *On the Origin of Species by Means of Natural Selection, or the Preservation of Favored Races in the Struggle for Life.*'
Alisha said, 'the 'preservation of favored races' part sounds a bit off, doesn't it?'
'To us, today, yes. But don't be too hard on Darwin, he was of his arrogant, imperialist, Victorian time and regarded *his* 'race' as favored and as the best: "At some future period", Darwin wrote, "not very distant as measured by centuries, the civilized races of man will almost certainly exterminate, and replace, the savage races throughout the world."[281]

'Tragically, Hitler took kindly and well-meaning Darwin, who was merely musing philosophically (if unpleasantly by modern standards) at his word, and thought he knew how to 'assist' the process of selection of who were, in his opinion, the 'fittest'. Stalin was no better and wiped out millions in pursuit of creating a materialist–atheist paradise on earth. Hundreds of millions of people suffered appallingly. So much for, 'brave new worlds' and materialist freedom or paradise.

'Darwinian–materialism is not only a dangerous ideology but it's also an irrational one. It reduces all things to 'dead', Life–Forceless, mineral matter and accidental, zombie 'machines', Soulless, 'ghostless' Cartesian automata that supposedly arose, and as importantly, arise right now, ongoing clever second by ongoing clever second, by 'accidental' and 'purposeless' processes and, quite bizarrely, never seems to notice that its no-intelligence-needed, just *Accidental Chemicals + Chemical Accidents* 'explanations' of 'Life' not only do not explain Life, they do not even *describe* it. As Larry Dossey said:[282]

> 'Novelist Arthur Koestler poked fun at these positions by taking aim at Rene Descartes, the seventeenth-century philosopher who was extraordinarily influential in establishing the notion of a mindless body. "If ... Descartes ... had kept a poodle, the history of philosophy would have been different," Koestler wrote. "The poodle would have taught Descartes that contrary to his doctrine, animals are not machines, and hence the human body is not a machine, forever separated from the mind ..."'[283]

---

[281] C. Darwin,1896. *The Descent of Man and Selection in Relation to Sex; The Works of Charles Darwin*, D. Appleton and Company, New York (First edition by AMS Press, 1972), p. 242.

[282] Dossey, L. 2010. *Is the Universe Merely A Statistical Accident? The Blog. Huffington Post.*

[283] Koestler A. 1978. *Janus: A Summing Up.* Random House; p. 229.

'Atoms of Oxygen, Carbon and Hydrogen etc, smart-materials-chemicals as they are, *don't feel sad or fearful.They don't get angry, they don't enjoy or think or plan or dream.* They are, obviously, not *Alive* in the way Plants and Animals and People are. They don't need to *breathe or feed or excrete.* Compared to the Plants, Animals and People they are relatively inanimate. Anima, after all, means *Soul.*

'This is why, Alisha, our civilization needs to call time on *all* materialistic, *no-intelligence-Life-Force-or-Soul-needed* explanations of the eons-long creation-in-evolution or ELCIE of the Living World. They are not merely scientifically unreasonable they are scientifically impossible and philosophically unreasonable. To maintain otherwise is a delusion for which the Victorian materialists, with their minimal knowledge of the stunning, 24/7 complexities of bio-chemistry, might have had some excuse, but we have none.

'So much turns on this. After all, what is more important than knowing whether our existence is meaningless or *intelligent* and *purposeful,* is heading somewhere, is intelligently evolving, that there is a deeper plan unfolding, no matter how dimly sensed by most of us, so far, are its outlines, that we *do* have Souls, that *intelligent,* mental Life after 'death', and probably reincarnation for further *human* lives, after rest and assimilation in the spiritual realms of multi–dimensional existence really is true and so on.

'Some materialist–minded science popularizers very much use the unproven Darwinian theory of random evolution to try to prove that 'science *is* materialism' and 'materialism *is* scientific', and that collective humanity's countless intuitions, psychic, mystical and spiritual experiences, visions and revelations that Life and the Living World actually *mean* something, that there really *are spiritual* purposes to existence, are all just so much make believe. Science, however, ceases to be science when it becomes a belief system or ideology that is *unwilling to accommodate new information,* such as the modern evidence for intelligent design in evolution, for a holistic, multi-dimensional cosmos of intelligence and meaning, for extrasensory faculties, out of body experiences and for Life after death.

'So this is why, Ollie, we must challenge the Darwinian–materialist concept of evolution as an Emperor's Clothes style delusion.'

'Precisely. This is why more of us need to point out that the amazing unfoldment of the Living World, over vast time, cannot be *rationally* explained by reference to purely *material and chance causes.* That it's a deep, non-evidence-based delusion to claim such a thing. If we succeed, and people agree, it will help to reduce the emotional despair and spiritual carnage caused by those who ceaselessly argue that Life is a second by second by second 'accident', that there are no *higher* purposes to Life and Nature, no Soul or Spirit or Life After Death or any real meaning at all and that the best proof that this dismal, materialist view is true is the unproven Darwinian theory of unintelligent and 'accidental' evolution.'

'Is life an 'accident', Ollie, or is it a miracle?'

'Yes, that's it. Those, fundamentally, are *our only two choices*. The funny thing is, Alisha, I've come to see it is not really a matter of *science* at all as to which we opt for – *intelligent* and purposeful evolution or *accidental* and meaningless evolution – but one of heart, mind, mood and taste. Do we look at all that exists in a glass half-full or half-empty kind of way? Are we excessively skeptical and cynical and inclined to debunk the meaningful, the Soulful and the spiritual or are we full of joy, faith, trust and hope – funnily enough, none of these traits, positive or negative, are qualities of matter as such but of mind and *Soul*. These are the real drivers and determinants of our beliefs and intellectual positions on Life on Earth's amazing and magical unfoldment across vast time, *not 'science'*.

'After all, if it was a matter of science and logic alone *no one* would insist, even for a moment, in such positively self-refuting disregard of *all* our real-world knowledge of the true evolutionary powers of chemical accidents – rust, frost, corrosion, erosion, forest fires, randomly evolved river valleys – that random, Life–Forceless, Soulless, unintelligent processes could explain the blazingly clever unfoldment, across vast time, of the Soul–Life–Form Beings of the Plants, Animals and People, the smartest, 24/7–dynamically *Functional, Conscious, Aware, Sentient, Living, Breathing, Thinking, Feeling, Acting, Reproducing,* 'mechanisms' we know of. The materialist line that the mysterious and magical Living World is just some kind of weird and never-ending *'accidental chemical reaction'* is false, it's a delusion.'

'Let's hope, Oliver, we can find a way to tell others about this and wake everyone up. Children and young people need to be told the truth.'

'I agree, Ali, they do. Perhaps we'll find a way to do so.'

We walked on through the lush green meadows by the quietly flowing river.

CPSIA information can be obtained at www.ICGtesting.com
Printed in the USA
BVOW041417140513

320692BV00009B/168/P